A HANDBOOK OF
PHYSICS MEASUREMENTS

BY

ERVIN S. FERRY

IN COLLABORATION WITH

O. W. SILVEY, G. W. SHERMAN, Jr.
AND D. C. DUNCAN

VOL. I
FUNDAMENTAL MEASUREMENTS,
PROPERTIES OF MATTER AND OPTICS

FIRST EDITION

NEW YORK
JOHN WILEY & SONS, Inc.
London: CHAPMAN & HALL, Limited
1918

Stanhope Press
F. H. GILSON COMPANY
BOSTON, U.S.A.

PREFACE

As stated in the preface to one of the earlier forms in which a part of this book has appeared:

"The aim of the present work is to furnish the student of pure or applied science with a self-contained manual of the theory and manipulation of those measurements in physics which bear most directly upon his subsequent work in other departments of study and upon his future professional career.

"Only those experimental methods have been included that are strictly scientific and that can be depended upon to give good results in the hands of the average student. Although several pieces of apparatus, experimental methods, and derivations of formulæ that possess some novelty appear, our fixed purpose has been to use the standard forms except in cases where an extended trial in large classes has demonstrated the superiority of the proposed innovation.

"It has been assumed that the experiment is rare that should be performed before the student understands the theory involved and the derivation of the formula required. Consequently the theory of each experiment is given in detail and the required formula developed at length. The more important sources of error are pointed out, and means are indicated by which these errors may be minimized or accounted for.

The book is designed to be commenced during the second college year. The greater part of the experiments requires no mathematics beyond trigonometry and college algebra. But wherever the calculus methods would result in economy of time and mental effort they have been employed.

No student except one specializing in physics would perform all 108 experiments included in the two volumes. Other students, after performing the necessary experiments on the properties of matter, would limit themselves to the groups bearing directly upon

iii

their principal study. Thus, few technical students except those of electrical engineering would do the work on damped vibration and harmonic analysis; few except those of chemical engineering would do the work on indices of refraction, polarimetry, and quantitative spectrum analysis.

ERVIN S. FERRY.

PURDUE UNIVERSITY, LAFAYETTE, INDIANA,
         June 14, 1918.

CONTENTS

VOLUME I

CHAPTER I

GENERAL NOTIONS REGARDING PHYSICS MEASUREMENTS

CHAPTER II

FUNDAMENTAL MEASUREMENTS AND THE PROPERTIES OF MATTER

CONTENTS

CHAPTER III

OPTICS

TABLES

A HANDBOOK OF
PHYSICS MEASUREMENTS

CHAPTER I

GENERAL NOTIONS REGARDING PHYSICS MEASUREMENTS

1. Introductory. — Experimental work has one of two objects; either to find out *what kind* of a result follows under given conditions, or to find out the *numerical relations* between different quantities. The first class of experiments is called *qualitative*, the second *quantitative*. In the earlier days of any science qualitative experiments are numerous; when the science is more mature, the majority of the experiments are quantitative. The determination of various quantitative relations is the object of physics measurement.

In making a physics measurement, the magnitude of each quantity concerned has to be expressed in terms of some *unit*, and the process of measurement consists essentially in finding how many times this unit is contained in the given quantity. The distance between two points, for example, may be expressed in terms of the number of foot rules which could be laid end to end between those points.

Some quantities can thus be measured *directly*, others can be measured only *indirectly*. Thus the Young's modulus of a brass wire cannot be experimentally determined by finding how many times the unit of Young's modulus is contained in the Young's modulus of the wire. The Young's modulus of the wire is usually determined by measuring a force and three lengths, and from

1

them calculating the Young's modulus. The great majority of physics measurements are indirect measurements.

2. Errors. — Every measurement is subject to errors. In the simple case of measuring the distance between two points by means of a steel scale, a number of measurements usually give different results, especially if the distance is several meters and the measurements are made to small fractions of a millimeter. The errors introduced are due in part to:

(1) Inaccuracy of setting at the starting point;

(2) Inaccuracy of setting at intermediate points when the distance exceeds one meter;

(3) Inaccuracy in estimating the fraction of a division at the end point;

(4) Parallax in reading, *i.e.*, the line from the eye to the division read not being perpendicular to the scale;

(5) The steel scale not being straight;

(6) The temperature not being that for which the steel scale was graduated;

(7) Irregular spacing of divisions;

(8) Errors in the standard from which the division of the steel scale was copied.

Besides the above there are doubtless other sources of error. It may be well here to note that blunders, such as mistakes due to mental confusion in putting down a wrong reading, or mistakes in making an addition, are not usually classed as errors.

Of the above errors, (1), (2), and (3) can be very much decreased by having fine divisions on the scale and reading with microscopes; (4) can be made small by bringing the scale on the steel scale close to the object to be measured; (5) can be made very small by using a steel scale of special design, or, in rough work, by holding the steel scale against a straight edge; (6) can be nearly eliminated by using the steel scale only at the proper temperature, or, if its temperature and coefficient of expansion are known, by calculating a *correction* to be applied; (7) can be diminished only by a careful comparison of the lengths of the different divisions; and for (8) corrections can be applied only when something is known about the accuracy of the standard from which the steel scale was copied. But even with the most refined methods and the most careful application of corrections, different measurements of the same distance usually give different results.

Errors due to (6), (7), and (8) may be *determinate* errors, *i.e.,* errors for which more or less accurate corrections can be calculated, whereas those due to (1), (2), and (3) are *indeterminate* errors, *i.e.,* errors for which corrections cannot be calculated. Moreover, of those errors for which corrections are not applied, some, like those due to (1), (2), and (3), will be *variable* in amount and will tend to make the value obtained sometimes too large and sometimes too small; while others, like those due to (7) and (8) when corrections for them are not applied, will be *constant* and will tend to make the value obtained always too large or always too small.

Since the average value of those variable errors which tend to make a result too large will after a considerable number of measurements be about the same as the average value of those variable errors which tend to make the result too small, the mean of a large number of measurements is usually nearly free from variable errors. In order as nearly as possible to do away with constant errors, the same quantity should be measured by as many different methods as possible. The results by the different methods will usually differ somewhat, but from them all a value can be calculated which is probably nearer the true value than any one of the separate results.

The magnitude of an error may be defined as the amount by which the value obtained exceeds the true value. That is, if the true value — which is not usually known — is denoted by T, the value obtained by O, and the error by E,

$$E = O - T. \tag{1}$$

The magnitude of the correction which ought to be applied may be defined as the amount which would have to be added to the value obtained in order to get the true value. That is, if C denotes the required correction,

$$C = T - O. \tag{2}$$

From (1) and (2) it will be seen that the error in a measurement and the correction which ought to be applied to it are equal in magnitude and opposite in sign. This does not mean that the error is exactly equal in magnitude to a correction which actually is applied, because for the correction itself only an approximate value is usually known.

3. Trustworthy Figures. — Since all measurements are subject to errors, it is important to be able to determine how many figures of a result can be trusted.

In direct measurements it is usually possible to make a fairly accurate estimate of the extent to which a reading can be trusted. Thus in reading by the unaided eye the position of a fine line which crosses a steel scale, the reading will not be in error by so much as a millimeter but pretty surely will be in error by more than a thousandth of a millimeter. So the extent to which the reading can be trusted will lie between these limits. A person who is accustomed to estimating fractions of a small division will be rather sure of not making an error so great as the tenth of a millimeter, and he can often trust his reading to a twentieth of a millimeter.

It is convenient always to put down all the figures that can be trusted, even if some of them are ciphers. Thus the statement that a distance is 50 cm. implies that there is reason for supposing that the distance really lies between 45 cm. and 55 cm., whereas the statement that the distance is 50.00 cm. implies that there is reason for supposing that the distance really lies between 49.95 cm. and 50.05 cm. When the distance is said to be 50 cm. the second figure is the last in which any confidence can be placed; when the distance is said to be 50.00 cm., the fourth figure is the last in which any confidence can be placed. If a distance is about 50,000 Km. and the third figure is the last in which any confidence can be placed, this fact may be indicated by saying that the distance is $50.0 \cdot 10^3$ Km.

In indirect measurements the result is usually calculated by some formula. To find out how many figures should be kept in the result consider the following two cases:

(a) If the result is the algebraic sum of several quantities, such as 314.428, 32.6, and 7.063, it is seen that in the sum, 354.091, no figure beyond that in the first decimal place can be trusted, because in the quantity which has the fewest trustworthy decimal places, viz., 32.6, no figure beyond the 6 can be trusted — otherwise it would have been expressed. So the sum will not be written 354.091, but 354.1. This suggests the following rule:

RULE I. — *In sums and differences no more decimal places should be retained than can be trusted in the quantity having fewest trustworthy decimal places.*

(b) If the result is the product of two quantities, such as 314.428 and 32.6, then the product cannot be trusted to more figures than appear in the quantity having fewest trustworthy figures, irrespective of the decimal place. To make this clear consider the following products:

$$314.428 \times 32.4 = 10187.4672$$
$$314.428 \times 32.6 = 10250.3528$$
$$314.428 \times 32.8 = 10313.2384$$
$$314 \times 32.6 = 10236.4$$

It is seen that if the quantity which is supposed to be 32.6 is really 32.4 or 32.8, then after the first three figures the true value of the product differs materially from the value obtained. The second and fourth of the above products show that if more than three figures cannot be trusted in one of two quantities which are to be multiplied, it is not worth while to use more than three — or at most four — figures of the other. These facts suggest the following rule:

RULE II. — *In products and quotients no more figures should be kept than can be trusted in the quantity having fewest trustworthy figures.*

Until the final result is reached, it is often worth while to keep one more figure than the above rules indicate.

For logarithms a safe rule is the following:

RULE III. — *When any of the quantities which are to be multiplied or divided can be trusted no closer than 0.01 per cent use a five-place table, when any of them can be trusted no closer than 0.1 per cent use a four-place table, and when any of them can be trusted no closer than 1 per cent use a slide rule.*

4. Required Accuracy of Measurement. — From Rule I it will be seen that if a small quantity is to be added to a large one, the percentage accuracy of the measurement of the small quantity need not be so great as that of the large one. Thus if $H = a + b$, and if a is about 100 cm. and b about 1 cm., a 1 per cent error in a will produce in H no greater effect than a 100 per cent error in b. When quantities are to be added or subtracted, they should be measured to the same number of decimal places.

From Rule II it will be seen that if a small quantity and a large one are to be multiplied the percentage accuracy of the measurement of the small quantity should be at least as great as that of the large one. Thus if $H = ab$, a 1 per cent error in a will produce in H the same effect as a 1 per cent error in b. So that if a is about 100 cm. and b about 1 cm., and if b cannot be trusted closer than 0.01 cm., there is no gain in accuracy by measuring a much closer than to within 1 cm. When quantities are to be multiplied or divided, they should be measured to within the same fraction of themselves, *e.g.*, all of them within 1 per cent and none of them much closer, or all of them within 0.01 per cent and none of them much closer.

The last statement needs modification in the case of a power. If the value found for a quantity a is 1 per cent too large, *i.e.*, is $1.01\,a$, then the value that will be obtained for a^2 is $1.0201\,a$, which is about 2 per cent too large, and the value obtained for a^3 is $1.030301\,a$, which is about 3 per cent too large. In general, if the value found for a is k per cent too large, the value that will be obtained for a^n will be nk per cent too large. So that a quantity which is to be squared, cubed, or raised to some higher power should be measured with more care than if it entered the formula to the first power.

ERRORS INTRODUCED BY COMMON APPROXIMATIONS

Number	True value	Approx. value	When applicable	How obtained	Approx. error introduced by the approximation
1	$1+a+a^2$	$1+a$	a small	Neglect a^2	$-a^2$ e.g. $\begin{cases} a & \text{error} \\ 0.1 & -\ 1\% \\ 0.01 & -\ 0.01\% \end{cases}$
2	$(1+a)(1+b)$	$1+a+b$	a and b small	Neglect ab	$-ab$
3	$(1+a)^m$	$1+ma$	a small	Expand by binomial theorem. Neglect second and higher powers of a	$-\dfrac{m(m-1)}{2}\cdot a^2$
4	$(1+a)^2$	$1+2a$	a small	Apply (3)	$-a^2$
5	$\dfrac{1}{1+a}$	$1-a$	a small	$\dfrac{1}{1+a}=(1+a)^{-1}$ Apply (3)	$-a^2$
6	$\sqrt{1+a}$	$1+\tfrac12 a$	a small	$\sqrt{1+a}=(1+a)^{\frac12}$ Apply (3)	$+\tfrac18 a^2$
7	$\sqrt{ab}$	$\tfrac12(a+b)$	b nearly equal to a	Let $b=a+e$. Then $\sqrt{ab}=\sqrt{a^2+ae}=a\sqrt{1+\dfrac{e}{a}}$ Apply (6)	$+\dfrac{(b-a)^2}{8a}$
8	$\sin a$	$a\,^*$	a small	$\sin a = a - \dfrac{a^3}{\underline{3}} + \dfrac{a^5}{\underline{5}} - \cdots$ Neglect third and higher powers	$+\tfrac16 a^3$
9	$\cos a$	1	a small	$\cos a = 1 - \dfrac{a^2}{\underline{2}} + \dfrac{a^4}{\underline{4}} - \cdots$ Neglect second and higher powers	$+\tfrac12 a^2$
10	$\tan a$	$a\,^*$	a small	$\tan a = \dfrac{\sin a}{\cos a} = \dfrac{a - \dfrac{a^3}{\underline{3}} + \cdots}{1 - \dfrac{a^2}{\underline{2}} + \cdots}$ Apply (5)	$-\tfrac13 a^3$
11	$\tan a$	$\sin a$	a small	Like (8) and (10)	$-\tfrac12 a^3$

* Expressed, of course, in radians.

5. Approximate Formulæ. — Beside the errors of observation, errors may be introduced into indirectly measured quantities by the use of formulæ which are only approximate. Thus, the sine and tangent of small angles are used as equal to the angles; the reciprocal of $(1 + a)$ is written equal to $(1 - a)$ when a is small; 3.14 is used for π; a number of figures are dropped from the end of a product, etc. Whenever such an approximation suggests itself, the error introduced by using it should be investigated and the approximation not made unless the error thereby introduced is so small as not to affect any figure that could otherwise be trusted in the result.

The preceding table of a few common approximations may prove useful.

6. Methods of Expressing Results. — The object of a quantitative experiment is sometimes the measurement of some quantity, and sometimes the determination of the relation between various quantities. When the relation between several quantities is sought, the usual method is, keeping all but two of the quantities constant, to vary by known amounts one of these two and then determine the changes produced in the other. Another pair of the quantities is then varied while the rest are kept constant, and so on until a sufficient number of pairs of quantities have been investigated. The various relations found to exist between the various pairs of quantities can then be combined to give the relation sought.

When one quantity has been given various known values and the corresponding values of a second quantity have been determined, the relation between them can always be expressed graphically; it can also be expressed more or less accurately by means of an empirical formula; and when this formula is sufficiently simple, the relation can without difficulty be expressed in words.

7. The Graphical Representation of Results. — Suppose that it is desired to determine the relation between the distance a body has fallen from rest and the time it has been falling. Suppose that a number of determinations are made, in each of which a ball is allowed to fall a known distance, and the time required is observed, the values obtained being those in the following table:

Distance fallen	Time required	Distance fallen	Time required
cm.	sec.	cm.	sec.
2.00	0.064	30.00	0.247
5.00	0.101	40.00	0.286
10.00	0.143	60.00	0.350
20.00	0.202		

These values may be plotted in the same way that curves are drawn in Analytic Geometry. The scales should be so chosen

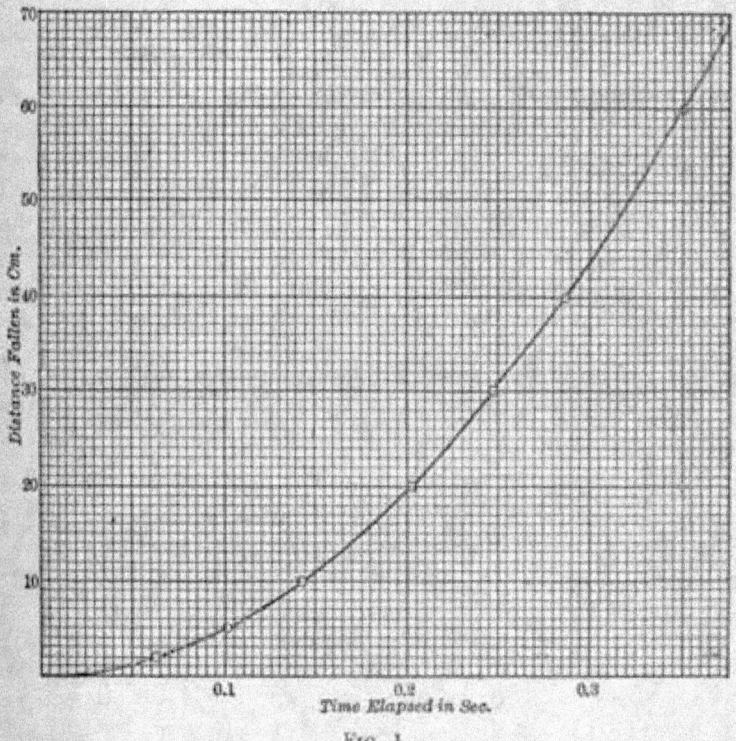

Fig. 1.

as to make the curve extend nearly across the sheet in both directions, unless by so doing a unit in the last place that can be trusted is represented by a distance greater than one of the smallest divisions of the paper. If, for instance, times can be trusted only to 0.01 sec., then the scale for abscissæ chosen in Fig. 1 is two or three times what it ought to be. If, however, the times can be trusted to 0.004 sec. or closer, then the scale is satisfactory. A curve should not always be drawn through all the points, but should be a smooth curve which fits the points as nearly as possible. If either scale has been so chosen that a unit in the last place that can be trusted is represented by one of the smallest divisions on the paper, a deviation of points from this curve usually indicates errors of observation.

The curve in Fig. 1 shows at once that the distance fallen increases as the time increases; but since the curve is not a straight line, the distance fallen is not proportional to the time. Since the curve is convex toward the time axis, it follows that the distance increases at a continually increasing rate, *i.e.*, that as the body falls it goes continually faster and faster. The curve also serves to find the distance fallen in any time not much exceeding 0.3 sec., or to find the time required to fall any distance not much greater than 60 cm.

The equation of this curve can be readily determined by the ordinary methods of analytic geometry.

8. Laboratory Rules and Procedure. — The work is divided into two parts: First, the experimental work; Second, the report on the experiment. These two are assigned to alternate meetings of the class.

In preparing for the work of the first period, study carefully the assigned references for the experiment scheduled for the day. Before coming to class, fasten with string, several sheets of Form "B" paper into a Laboratory Cover and fill out the blanks neatly in freehand lettering. This cover with enclosed sheets is to be submitted to the instructor in charge. The student will then write an analysis of the experiment. This is to include statement of the problem at hand, a definition of the physics quantities measured, with a complete connected story of the experiment in which is discussed the theory, physics principles involved, the mathematical discussion and a brief procedure. This is to be written under class-room conditions without use of books or notes. The first page is to be left blank for the instructor's corrections. Submit this analysis to the instructor for his approval.

Next proceed to obtain the necessary data according to directions. Record every reading made, even if it is exactly the same as the previous one. Before leaving the laboratory there is to be filed with the instructor a data sheet, Form "K," containing all the observations made, but not calculated values. No numerical value is to be given without a statement of the unit employed. Numbers of all important pieces of apparatus and position number should be recorded on this sheet.

On the second period all corrections on the analysis are to be made first, and then the calculations made from the data. All results should be expressed in both c.g.s. and f.p.s. system where possible. Discuss accuracy of results, by comparing with accuracy of individual readings. State any conclusions to be drawn from the experiment. Whenever possible, results are to be expressed in the form of a curve plotted on Form "F." This should be done with drawing instruments and India ink. Give the curve a title, e.g., "Calibration Curve of Hydrometer No. 26983."

Do not leave the laboratory until the report is accepted by the instructor in charge.

CHAPTER II

FUNDAMENTAL MEASUREMENTS AND THE PROPER-
TIES OF MATTER

9. The Measurement of Distance. — The vast majority of the measurements made in a physics laboratory are ultimately measurements of distance. Two temperatures, for instance, may be compared by the difference in the lengths of a thread of mercury; a pressure may be determined from the height of a barometric column, or from the distance that the pointer of a pressure gauge moves; a difference in time may be measured by the distance that the hand of a clock has moved; etc.

10. The Meter Stick. — This is the instrument most often used in the laboratory for the measurement of moderate distances. Usually the smallest divisions marked on it are millimeters. Since the last division at each end is liable in time to become worn a trifle short, the ends are seldom employed. In use, the meter stick is turned up on its side so as to bring its scale as close as possible to the object to be measured, some line on the meter stick is brought as nearly as possible into coincidence with one end of the distance to be measured, and the reading of each end of the distance is noted, the tenths of a millimeter being estimated. The difference between the two readings gives the distance sought. Division lines which are as close together as a fifth of a millimeter are usually more confusing than helpful. A very little practice, however, will make possible the rather accurate estimation of a tenth of a division, provided the division is not much smaller than a millimeter.

11. The Micrometer Screw. — For the more accurate measurements of small distances, the principle of the micrometer screw has many applications. A carefully made screw with a divided head turns in an accurately fitting nut. An index mark close to the divisions on the head shows through how many divisions the screw has turned. The distance between the threads of the screw divided by the number of divisions on the head gives the distance the end of the screw advances when the head is turned through one of its divisions. The principle of the micrometer screw is employed in the micrometer caliper, the spherometer, the dividing engine, and the filar micrometer microscope.

The *Micrometer Caliper* (Fig. 2) consists of an accurately made screw which can be advanced toward or away from the stop A. The whole number of millimeters distance between A and B is indicated by the millimeter divisions on the shank C uncovered by the sleeve D, while the fraction of a millimeter is given by the graduated circle on the edge of the

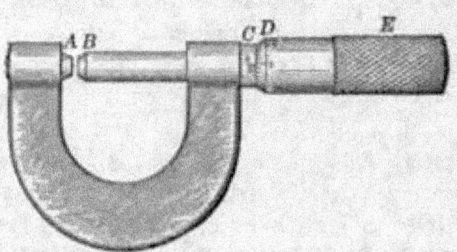

Fig. 2.

sleeve D. If the pitch of the screw is half a millimeter and if the head is divided into fifty equal spaces, one division on the shank will be uncovered by the sleeve for every two complete turns of the screw, and each space on the divided head corresponds to an advance of the screw of 0.01 mm. Thus if tenths of a division on the sleeve are carefully estimated, a reading can be trusted to 0.0005 mm.

The "zero reading" of the instrument, *i.e.*, the reading when B just touches A, should always be recorded. In making a reading, the sleeve is never turned up tight, but only until a very slight pressure is felt.

In the *Spherometer*, a micrometer screw passes vertically through a nut mounted at the center of an equilateral tripod. A distance corresponding to an advance of the screw through a fraction of a turn is indicated either by a large divided head attached to the screw and which moves past a pointer fixed to the nut, or by a pointer attached to the screw, Fig. 3, and which moves around a divided circular scale fixed to the nut. The pitch of the screw is frequently ½ mm. and the circular scale is divided into 500 equal spaces, so that by estimating tenths of a division, a reading can be made within 0.0001 mm. However, with most spherometers, several successive settings show that they cannot be trusted much closer than 0.001 mm., so that it is useless to read the fractions of a division. The spherometer is especially useful in measuring the radius of curvature of spherical surfaces — whence its name.

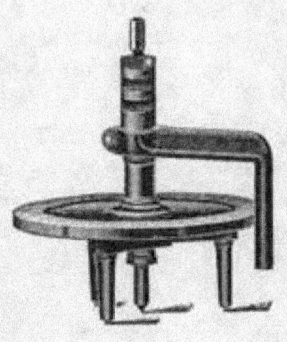

Fig. 3.

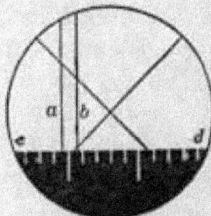

Fig. 4.

The *Filar Micrometer Microscope* is a microscope that has in the focal plane of the eyepiece two parallel cross hairs, *a* and *b* (Fig. 4), which can be moved across the field of the microscope by means of a micrometer screw. In the focal plane there is also a fixed serrated edge, *cd*, the teeth of which serve as a scale to indicate the whole number of turns made by the micrometer screw. The distance on the microscope stage corresponding to one turn of the micrometer screw must be determined by focalizing the microscope on a standard scale. The standard commonly used is a scale having ten divisions to the millimeter. Care is taken to have the lines of the standard scale parallel to the movable cross hairs. Readings are made on, say, five consecutive lines of the standard scale near the left side of the field of view, and then on the same number near the right side of the field. From the difference between the readings for the left-most lines of the two sets is obtained one determination of the distance corresponding to one turn of the screw; from the difference between the readings for the second lines in the two sets is obtained a second determination; and so on.

If the pitch of the screw is such that one turn corresponds to a distance of 0.1 mm. on the microscope stage, and if the head is divided into 50 parts, one division on the head corresponds to 0.002 mm. With the best microscope it is impossible to distinguish lines closer together than about 0.001 mm., but the mean of a number of careful settings on a very fine line can be trusted to about 0.0005 mm. In making a setting, the screw should always be turned up from the same direction in order to avoid errors due to backlash.

12. The Eyepiece Micrometer. — This is a finely divided scale ruled on thin glass placed in the focal plane of a microscope. The eyepiece micrometer is standardized as follows: Adjust the position of the ocular till the lines on the glass scale are distinct. Thereafter, the position of the ocular with respect to the glass scale must not be changed. Move the microscope up toward a standard scale till the image of the latter is sharp. Noting the number of divisions of the eyepiece scale that correspond to a given number of divisions of the standard scale, the value of a single division of the eyepiece scale is determined. This value will hold only when the object distance is the same as when the instrument was standardized. But if the distance between the ocular and eyepiece scale be fixed, the object distance will be the same whenever the object is in sharp focus.

13. The Vernier Scale. — Vernier's scale is a device for the estimation of fractions of the smallest divisions of a scale. It consists of a short auxiliary scale, called the "vernier," capable of sliding along the edge of the principal scale. The precision attainable with the vernier scale is about three times that attainable with the unaided eye.

Usually, the vernier is divided so that n divisions of the vernier correspond to $(n - 1)$ divisions of the main scale. In some verniers, however, n divisions of the vernier equal one less than some multiple of n divisions of the main scale. In either case, one nth of the length of a division of the main scale is called the *least count* of the vernier, where n represents the number of spaces into which the vernier is divided.

Thus, in the case of a vernier divided so that n spaces of the vernier correspond to $(n - 1)$ spaces on the main scale, if s be the length of the smallest spaces on the main scale, and v the length of the shortest divisions on the vernier,

$$nv = (n - 1) s,$$

or, the least count,

$$\frac{s}{n} = s - v. \tag{3}$$

The theory of the vernier may be made clear by the following example: Suppose that along a meter stick there slides a vernier 9 mm. long divided into ten equal parts. Each division on the vernier is then 0.9 mm. long, and

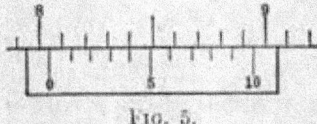

FIG. 5.

if the 0-mark and the 10-mark of the vernier coincide with lines on the meter stick, then the 1-line on the vernier lacks 0.1 mm. of coinciding with a line on the meter stick, then the 2-line lacks 0.2 mm. of coinciding with a line, the 3-line lacks 0.3 mm., and so on. If, then, the vernier were to be moved along 0.3 mm., its 0-line would be 0.3 mm. beyond some mark on the meter stick, and the 3-line would coincide with some mark; if the 7-line coincided with some mark, the 0-line would be 0.7 mm. beyond some mark, etc. The position of the 0-line is what is desired. In Fig. 5 the reading is 8.04 cm.

In the case of a vernier divided so that n spaces of the vernier equal one less than some multiple a of n spaces on the main scale we have

$$nv = (an - 1) s.$$

Whence, the least count is

$$\frac{s}{n} = as - v. \tag{4}$$

In using any vernier scale we first bring the first and last lines of the vernier into coincidence with lines of the main scale, and then note the number of divisions in the given space as indicated on the vernier scale and on the main scale. From this observation we obtain the least count of the vernier. The least count multiplied by the number of the vernier line which coincides with a line on the scale gives the distance between the 0-line of the vernier and the preceding line on the scale. In the case of a circular scale divided into thirds of a degree, the vernier is often made fifty-nine thirds of a degree long and is divided into sixty equal parts. Its least count is then one third of a minute. Fig. 6 shows such a vernier, and also illustrates the manner in which verniers are often numbered so that readings can be made directly without computa-

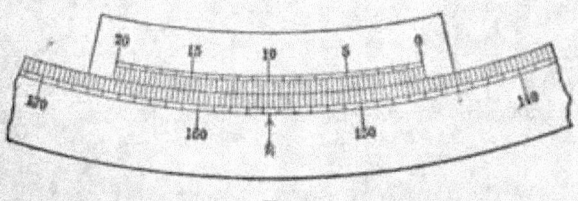

Fig. 6.

tion. In this particular case, since each vernier division corresponds to one third of a minute, it is natural to number the fifteenth division 5, the thirtieth division 10, etc., minutes. The reading is 145° 50′ 0″.

The *Vernier Caliper* (Fig. 7) consists of a finely divided steel scale C with a fixed jaw at one end, and a jaw B provided with a

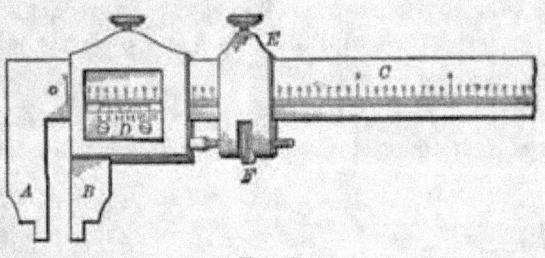

Fig. 7.

vernier scale D that can slide along the length of the scale. In using this instrument the jaw B is nearly closed upon the object to be measured, the screw E is tightened, and the final adjustment carefully made with the screw F. The zero reading should always be noted, and care should be taken that F is turned only until a slight pressure is felt.

14. The Cathetometer.—The cathetometer is an instrument for measuring vertical distances in cases where a scale cannot be placed very close to the points whose distance apart is desired. It con-

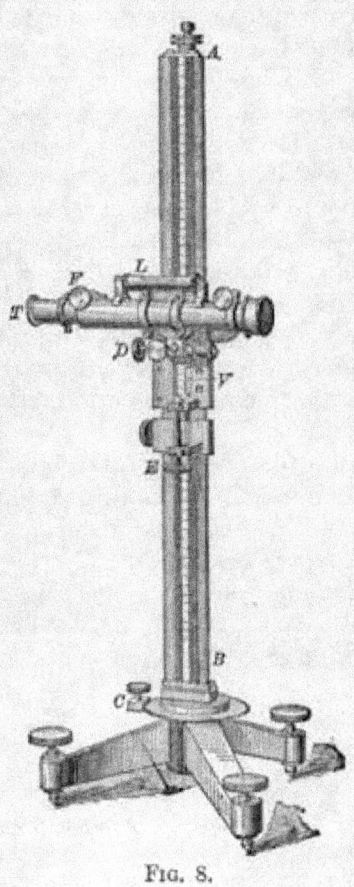

sists essentially of an accurately graduated scale, Fig. 8, together with a horizontal telescope capable of being moved up and down a rigid vertical column. The position of the telescope can be read off the scale by means of a vernier. In measuring the vertical distance between two points, the instrument must first be adjusted as described in the following paragraphs. Then the cross hair in the eyepiece of the horizontal telescope is brought into coincidence with the image of one point and the position of the telescope noted; the cross hair is then brought into coincidence with the image of the other point and the new position of the telescope noted. The difference between these readings is the vertical distance required.

Before taking a reading with a cathetometer three adjustments are necessary. The first adjustment is to make the axis AB vertical. To effect this, the telescope is set approximately parallel to the line connecting two of the three leveling screws in the base, and one or both of these two screws is turned until the bubble in L is near the middle of the vial. The telescope is then rotated about AB until it points in the opposite direction. If the bubble is not still in the middle, it is brought back to the middle by turning one

Fig. 8.

or both of the two screws, the number of turns being counted. Half of that number of turns is then made in the opposite direction and the bubble brought back to the middle of the vial by means of the screw D. The telescope is then turned so as to be 90° from its original position and the third screw in the base adjusted until the bubble is in the middle. If the bubble does not now remain in the middle of the vial, however the telescope may be turned about AB, the entire adjustment is repeated.

The second adjustment is to make the axis of the telescope horizontal. In doing this the telescope is taken from its wyes, turned end for end, and replaced. If the bubble does not come to rest at the middle of the vial, it is brought to the middle by the screw D, the number of turns required being counted. Half this number of turns is made in the opposite direction and the bubble then brought to the middle by means of the screws at the ends of the vial. The telescope is again reversed in the wyes, and if the bubble does not still come to rest in the middle, the above operations are repeated.

The third adjustment is to focalize the telescope. The front tube containing the eyepiece is moved in and out until the cross hairs appear as distinct as possible. Then, while sighting along the outside of the telescope, the latter is brought to about the right height and turned so as to point approximately at the object to be viewed. The eye is then placed at the eyepiece and the focalizing screw F turned until the image of the object does not move with reference to the cross hairs when the observer's head is moved slightly from side to side.

The cathetometer is now in adjustment. In finding the vertical distance between two points, the telescope is focalized first on one of the points and then on the other, the final setting being made in each case by the screw E. After each setting the height of a mark on the carriage is read by the vernier V. The difference between the two readings gives the desired distance.

15. Amsler's Planimeter. — A planimeter is a direct-reading instrument used to determine the areas of irregular figures on drawings. Amsler's polar planimeter (Fig. 9) consists of two arms, — a tracer arm AC, and a pole arm EC, — jointed at C.

The point E is fixed and the point A is carried around the boundary of the figure in the direction of the hands of a watch. Attached to

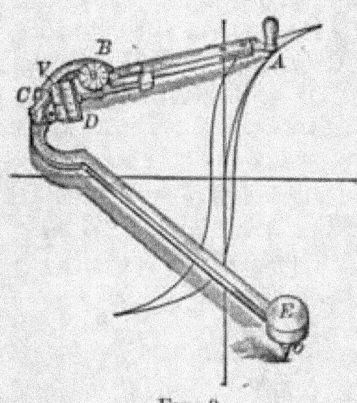

the tracer arm is a small roller D, the axis of which is parallel to the line AC. This roller and the points A and E are the only parts of the planimeter that touch the paper. As the point A passes over the boundary of the figure, the roller rotates unless the motion A is entirely in the direction of AC — in which case the roller slides. It will now be proved that when the tracing point circumscribes any closed plane figure which does not contain the pole point, the circumference of the roller rotates a distance proportional to the area circumscribed by the tracing point. This proof will be in four parts.

Fig. 9.

First, consider two concentric circular arcs AA'' and $A'A'''$ (Fig. 10) cut by radii AE and $A''E$. Let the pole point of the planimeter be fixed at the center of these arcs, and let the tracing point be moved along the radius AE from

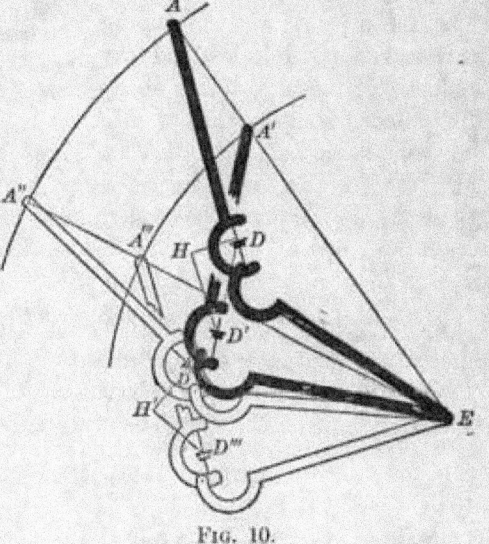

Fig. 10.

A to A'. Then the roller will move from D to D' while a point in its circumference will rotate through the distance DH. Again, let the tracing point be moved along the radius $A''E$ from A'' to A''', causing the roller to move from D'' to D''' while a point on its circumference rotates through the distance $D''H'$. Since the shape and size of the figure $ED'HDA'A$ are the same whether the tracing point has moved along the radius AE or along some other radius $A''E$, it follows that $D''H'$ equals DH. Therefore, while the tracing point passes over the portions of any radii intercepted between the same two circles having the pole point E as center, the rolling components of the motion of the roller are equal.

Second, let ECA (Fig. 11) represent the planimeter in one position, and $EC'A'$ the planimeter in another position. Draw EB and JD normal to AC, and HD' normal to JD; also draw EA, ED, and ED'. For brevity let δ_x denote the distance through which a point in the circumference of the roller moves with reference to AC when A describes any line x.

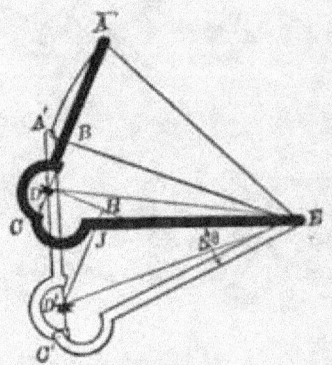

Let the instrument start from the position ECA, and, keeping the angle ECA constant, rotate about E through a small angle $\Delta\Theta$ into the new position $EC'A'$. AA' is, then, the arc of a circle described about E as center. The roller,

Fig. 11.

meantime, moves through a small distance DD', sliding through a distance HD', and rolling through a distance DH. Whence,

$$\delta_{AA'} = DH = DD' \cdot \cos HDD' = ED \cdot \Delta\Theta \cdot \cos HDD'. \quad (5)$$

Since HD is by construction normal to AC, and the very small arc DD' is normal to the radius ED, the angle HDD' equals the angle BDE. And since BE is by construction normal to BD,

$$\cos HDD' [= \cos BDE] = \frac{BD}{ED}.$$

Equation (5) becomes, therefore,

$$\delta_{AA'} = ED \cdot \Delta\Theta \cdot \frac{BD}{ED} = \Delta\Theta \cdot BD. \quad (6)$$

Now　　　$BD = BC - DC = EC \cdot \cos ACE - DC.$　　　(7)

And since in the triangle ACE,

$$(AE)^2 = (AC)^2 + (EC)^2 - 2\, AC \cdot EC \cdot \cos ACE,$$

(7) may be written

$$BD = EC \cdot \frac{(AC)^2 + (EC)^2 - (AE)^2}{2\, AC \cdot EC} - DC.$$

Equation (6) becomes, therefore,

$$\delta_{AA'} = \Delta\Theta \left[\frac{(AC)^2 + (EC)^2 - (AE)^2}{2\, AC} - DC \right].$$　　　(8)

For this particular case, then, where the tracing point moves over the very small arc of a circle described about the pole point,

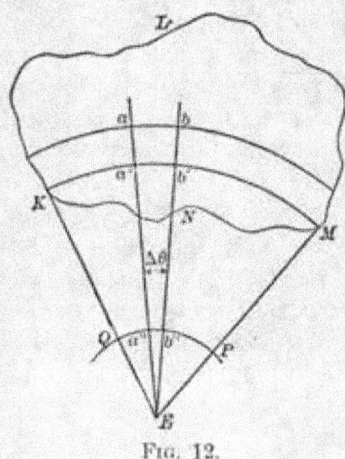

$\delta_{AA'}$ is expressed in terms of the radius of this circle, the dimensions of the instrument, and the very small angle subtended by the given arc.

Third, let any figure $KLMN$ (Fig. 12), not inclosing the pole point, be cut into a large number of very narrow strips by a series of circles having E for center. Let these strips be cut into very small areas by radii drawn from E. Thus the entire figure is divided into a great number of areas, each as small as we choose.

Fig. 12.

If the tracing point circumscribes in the clockwise direction one of these small areas, ab', we have, from the first division of this proof,

$$\delta_{bb'} = -\delta_{a'a}.$$

And since δ_{ab} is described in the opposite direction from that assumed in the second division of this proof, the entire value of $\delta_{ab'b'a'a}$ is equal to $\delta_{b'a'} - \delta_{ab}$. Whence, by (8),

$$\delta_{abb'a'a} = \Delta\Theta\left[\frac{(AC)^2 + (EC)^2 - (a'E)^2}{2\,AC} - DC\right]$$

$$- \Delta\Theta\left[\frac{(AC)^2 + (EC)^2 - (aE)^2}{2\,AC} - DC\right]$$

$$= \frac{\Delta\Theta}{2\,AC}[(aE)^2 - (a'E)^2].$$

But $\quad \frac{1}{2}\Delta\Theta\,(aE)^2 = \frac{1}{2}\Delta\Theta\,(aE)\,(aE) = \frac{1}{2}(ab)\,(aE),$

and this last expression measures the area of the circular sector abE. .In the same way $\frac{1}{2}\Delta\Theta\,(a'E)^2$ measures the area of the circular sector $a'b'E$. So that

$$\delta_{abb'a'a} = \frac{\text{area }(abE) - \text{area }(a'b'E)}{AC} = \frac{\text{area }(ab')}{AC}. \qquad (9)$$

In Fig. 11 the angle BDE was acute. If this angle be obtuse, it will be seen that the roller then rotates in the opposite direction. Calling rotation in this opposite direction negative, and making the changes in sign involved in the new figure, we find that (9) holds whether BDE is acute or obtuse. That is, when the elementary area ab' is circumscribed by the tracing point, that area is given by the product of the length of the tracer arm and the small distance through which a point in the circumference of the roller has rotated.

Fourth, let the tracing point move over the whole figure *KLMN* (Fig. 12) in such a way as to traverse the boundary once in a clockwise direction, and each of the radial lines and circular arcs twice, once in each direction. By taking lines in the proper order, this can be done without lifting the tracing point from the paper. Describing these lines in the manner indicated amounts to the same thing as going once in the clockwise direction around the whole figure; it also amounts to the same thing as going once in the clockwise direction around each of the small areas into which the figure is divided. The total value of δ_z will then be

$$\delta_{KLMNK} = \sum \frac{\text{area }(ab')}{AC} = \frac{\text{area }(KLMN)}{AC}. \qquad (10)$$

This equation shows that *when an area which does not contain the pole point is circumscribed by the tracing point, the area is measured by the product of the length of the tracer arm and the distance through which a point in the circumference of the roller has rotated with reference to the tracer arm.*

The dimensions of the planimeter are usually so selected that the product of the length of the tracer arm by the circumference of the roller is equal to ten square inches or a hundred square centimeters. That is, they are so selected that if the tracing point circumscribes an area of ten square inches or a hundred square centimeters, as the case may be, the roller rotates once. The circumference of the roller is then divided into a hundred equal parts, and these by means of a vernier (V, Fig. 9) can be read to tenths.

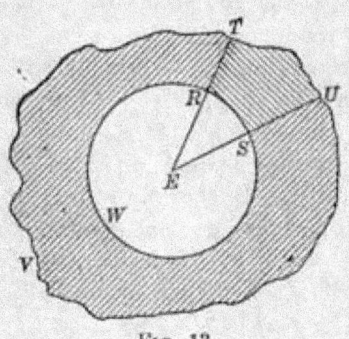

Fig. 13.

The counting wheel B indicates the whole number of revolutions of the roller.

In the practical use of a planimeter, the figure the area of which is desired may be so large that it cannot be circumscribed without placing the pole point inside it. In this case the area may be determined as follows:

If the angle ADE (Fig. 11) is a right angle, then BD is zero, and, therefore, from (6), $\delta_{AA'}$ equals zero. That is, as A moves about E in the circular arc AA', the roller slides, without rolling at all. The circle generated by the tracing point A about the pole point E as center when the roller does not rotate, and so makes no record, is called the "zero" or "datum" circle.

In Fig. 13, let RSW be this datum circle. Then if the tracing point were to circumscribe the area TS, (9) shows that

$$\delta_{TUSRT} = \frac{\text{area } (TS)}{AC}, \tag{11}$$

and if now the tracing point were to circumscribe the rest of the shaded area, then

$$\delta_{TRWSUVT} = \frac{\text{shaded area } (UVT)}{AC}. \tag{12}$$

If these two paths were to be described successively, then, by adding (11) and (12), we find that

$$\delta_{TUSRTRWSUVT} = \frac{\text{total shaded area}}{AC}.$$

In tracing this whole path, the lines US and RT have each been described twice, once in each direction, so that the resultant motion of the roller produced by tracing them is zero, and, since RSW is the datum circle, the roller did not rotate while it was traced. It follows that if the tracing point had simply described the perimeter $TUVT$, the roller would in the end have turned from its first position just as much as it did while the more complicated outline was being traced. That is, if the tracing point were to describe the entire perimeter of the figure, the area indicated by the roller would be the area of that part of the figure outside of the datum circle. If the tracing point were ever to cross the boundary of the datum circle, the roller would move in opposite directions before and after crossing. From this it may be shown, if proper attention be paid to the sign of the roller reading, *that whenever the pole point is inside of the figure circumscribed by the tracing point, the area actually circumscribed is greater than that indicated by the roller, the difference being the area of the datum circle.*

To sum up, *if the tracing point circumscribe in the clockwise direction any area, the difference between the final and initial readings of the roller gives the area when the pole point lies outside the figure; when the pole point lies inside the figure, the area is obtained by adding to this difference the area of the datum circle.*

16. The Beam Balance. — One of the most common as well as the most accurate methods for the comparison of masses is afforded by the beam balance (Fig. 14). The beam BB' can rotate about a knife edge K, which rests upon an agate plate. Suspended from knife edges at K_1 and K_2 are the scale pans p_1 and p_2. A handle H operates an arrestment consisting of a horizontal rod and three cams C_1, C_2, and C_3, by means of which the knife edges may be relieved of the weight of the beam and pans when the balance is not in use and when the masses in the pans are being changed. Fastened to the beam is a long pointer I which swings

in front of a graduated scale S.   Whether the divisions on this
scale are numbered or not, it is convenient to assume that the
middle division is numbered 10, and that the divisions are numbered from left to right.   Projecting from the side of the case is a
rod R by means of which a bent aluminium wire called a *rider* can
be placed at any point along the beam.   This rider is used in place
of standard masses smaller than 10 mg.   The top of the beam is
often divided into twenty equal parts, the 0-line being over the

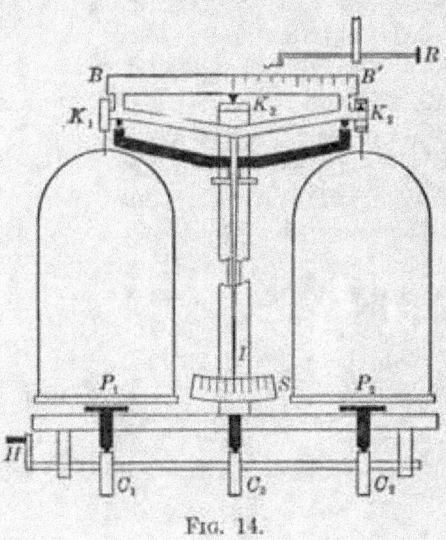

Fig. 14.

central knife edge, and the 10-lines over the other knife edges.   If
the mass of the rider is 10 mg., and it is placed on one of the 10-
lines, it produces the same effect as if a 10 mg. mass were in the
corresponding pan; but if it is placed at division 3, it has a turning
moment only three-tenths as great, and so produces the same
effect as would a 3 mg. mass placed in the pan.   Occasionally
a rider of some other mass is used and the beam divided accordingly.

17. The Method of Vibrations. — This is the method usually
employed in making accurate weighings.   When using this method,
the balance case is at first left closed and the arrestment released.

If the pointer does not begin to swing, the case is opened, the hand waved lightly over one pan, and the case again closed. With the pointer swinging in front of the scale, but not beyond it, the zero point of the balance is determined; *i.e.*, the point at which the pointer would finally come to rest, either with no load on the pans, or with equal loads on the two pans.

This is done by observing an odd number of successive turning points of the pointer. As the pointer swings, the distance between any two successive turning points on the same side of the scale gradually decreases, but in a few swings the decrease is slight. The zero point is about halfway from b (Fig. 15) to a point midway between a and c. It is also about halfway from c to a point midway between b and d; about halfway from d to a point midway between c and e; etc. Since the distance from a to c is about the same as that from c to e, the average of $a, c,$ and e is nearly the same as c. The zero point, then, is very near the point found by

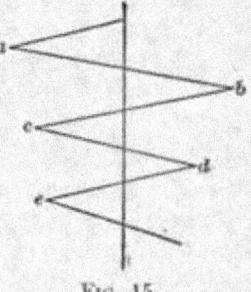

Fig. 15.

taking the average of $a, c,$ and e, and averaging with it the average of b and d.

Suppose, for instance, that five successive turning points are observed to be:

$$
\begin{array}{ll}
8.4 & \\
 & 11.9 \\
8.5 & \\
 & 11.8 \\
8.7 &
\end{array}
$$

Then the average of the turning points at the left is 8.53 and of the turning points at the right is 11.85. Consequently the zero point is in the neighborhood of $[\frac{1}{2} (8.53 + 11.85) =] 10.2$.

Five successive turning points are usually enough to observe. But any odd number of successive turning points may be used in the same way; *viz.*, by averaging the left turning points and averaging the right turning points, and then finding the average of the two results. It should be noted that this method of finding the zero point is most accurate when the pointer swings with a small amplitude. Since the zero point varies from day to day, and even from hour to hour, it should be determined for each experiment. For very accurate work it should be determined both at the beginning and at the end of a weighing, and the average value used.

After the zero point has been determined and while the arrestment is elevated so as to lift the beam off the knife edge, the object is placed on one pan and standard masses on the other. Right-handed persons find it most convenient to place the object on the left pan so that the mass pan is in front of the hand that makes the adjustment of the standards. Each time that a new mass is placed on the pan the arrestment is lowered just enough to see in which direction the pointer would swing, but no masses are ever put on or taken off while the pointer is free to swing. When the masses are so nearly adjusted that if the arrestment is entirely released the pointer swings back and forth near the zero point, the position at which the pointer would finally come to rest is determined from several successive turning points in the same way that the zero point had been. The rider is then moved so as to alter the effective mass on the mass pan by one or two milligrams, and the new position of rest determined. From these observations the mass which would be required to make the point of rest coincide with the zero point can be calculated without taking the time to effect the balance experimentally.

Suppose, for example, that the zero point of the balance is 10.2 scale divisions, and that with the object on the left pan and a mass of 24.166 g. on the right pan, the point of rest is found to be 11.6 scale divisions. Since this point of rest is to the right of the zero point, the mass on the right pan is too small. Suppose that by means of the rider the effective mass on the right pan is increased by 2 mg., and that the new point of rest, determined as before, is found to be 7.4 scale divisions. Then the addition of 2 mg. has moved the point of rest through [11.6 − 7.4 =] 4.2 scale divisions, and 1 mg. would have moved it 2.1 scale divisions. It follows that the mass which would have to be added in order to move the point of rest through the [11.6 − 10.2 =] 1.4 scale divisions to the zero point of the balance is [1.4 ÷ 2.1 =] 0.7 mg. Consequently the apparent * mass of the object is [24.166 + 0.0007 =] 24.1667 g.

18. Sensitivity. — The sensitivity of a balance is defined as the number of scale divisions through which the point of rest is

* See below, *Errors in Weighing.*

moved by the addition of one milligram to the load on one of the pans. In the above example the sensitivity was 2.1 scale divisions per milligram. The sensitivity, however, depends upon the load and should therefore be determined for each weighing. The fact that it depends upon the load may be shown as follows:

Let K_1, K_2, K_3 (Fig. 16) denote the three knife edges of the balance, and M the center of mass of the beam. Let p_1 and p_2 be the respective masses of the left and right pans, and M_3 the mass of the beam. Suppose that with a mass M_1 on the left pan and a mass M_2 on the right pan the beam comes to rest in the position

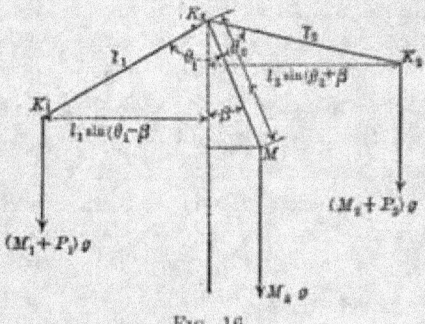

Fig. 16.

indicated. Then, since the balance is in equilibrium, the sum of the moments of $(M_1 + p_1) g$, $(M_2 + p_2) g$, and $M_3 g$ taken about K_2 must equal zero, i.e.,

$$(M_1 + p_1) g \times l_1 \sin (\theta_1 - \beta)$$
$$- (M_2 + p_2) g \times l_2 \sin (\theta_2 + \beta) - M_3 g \times r \sin \beta = 0,$$

or
$$(M_1 + p_1) l_1 (\sin \theta_1 \cos \beta - \cos \theta_1 \sin \beta)$$
$$- (M_2 + p_2) l_2 (\sin \theta_2 \cos \beta + \cos \theta_2 \sin \beta)$$
$$- M_3 r \sin \beta = 0. \qquad (13)$$

Since in the actual case β is always small, we may replace $\sin \beta$ by β, and $\cos \beta$ by 1. Then if $l_2 = l_1 = l$, if $\theta_2 = \theta_1 = \theta$, and if $p_2 = p_1 = p$, we have

$$(M_1 + p) l (\sin \theta - \beta \cos \theta)$$
$$- (M_2 + p) l(\sin \theta + \beta \cos \theta) - M_3 r \beta = 0.$$

Whence,

$$\frac{\beta}{M_1 - M_2} = \frac{l \sin \theta}{(M_1 + M_2 + 2p) l \cos \theta + M_3 r}. \tag{14}$$

If $M_1 - M_2 = 0$, then $\beta = 0$; and if $M_1 - M_2 = 1$ mg., then the left member of (14) denotes the movement of the pointer for 1 mg. change in the load. That is, the sensitivity of the balance is proportional to each member of this equation.

If θ be 90°, $\cos \theta$ is zero, and whatever the value of the load, $M_1 + M_2$, the right member of (14) is unaltered. That is, when θ is 90°, the sensitivity is independent of the load. If θ be less than 90°, $\cos \theta$ is positive, and as the load, $M_1 + M_2$, increases, the right member of (14) decreases. That is, when θ is less than 90°, the sensitivity decreases as the load increases. If θ be larger than 90°, $\cos \theta$ is negative. It follows that as $M_1 + M_2$ increases, the denominator of the right member of (14) decreases, and the sensitivity therefore increases. That is, when θ is larger than 90°, the sensitivity increases as the load increases. Since different loads necessarily bend the beam different amounts, it follows that the sensitivity is different for different loads. The maker usually arranges to have the three knife edges in line when the balance has about half its maximum load.

19. Errors in Weighing. — The errors to which a weighing is especially liable are due to (a) the buoyant effect of the air, (b) errors in the standard masses, (c) difference in the lengths of the balance arms, and (d) difference in the masses of the scale pans.

(a) The buoyant effect of the air will be different upon the bodies on the two scale pans unless their volumes are equal. The true mass may be found as follows: Let M, D, and V denote respectively the mass, density, and volume of the body the mass of which is desired, and m, d, and v, the mass, density, and volume of the standard masses which just balance it in air of density ρ. Then the difference between the weight of the body in vacuum and its weight in air is equal to the weight of the air displaced $\rho V g$, and the weight of the body in air is consequently $Mg - \rho V g$. In the same manner, the standard masses when in air weigh $mg - \rho v g$.

Since the weight of the body in air equals the weight of the standard masses in air,

$$Mg - \rho Vg = mg - \rho vg,$$

or

$$Mg - \rho \frac{M}{D}g = mg - \rho \frac{m}{d}g,$$

so that

$$M\left(1 - \frac{\rho}{D}\right) = m - \rho \frac{m}{d}. \tag{15}$$

(c) and (d). Errors due to difference in the lengths of the balance arms and to difference in the masses of the scale pans can be nearly eliminated by weighing the body first in one pan and then in the other.   This is called the Method of Double Weighing and will now be explained.

Let l_1 and l_2 denote the respective lengths of the left and right arms of the balance, and p_1 and p_2 the respective masses of the left and right pans.   If an object of mass M is balanced by standard masses m_1 when the object is in the right pan; and by standard masses m_2 when the object is in the left pan, then in Fig. 16, $\beta = 0$. If, in addition, $\theta_2 = \theta_1$, we have from the principle of moments,

$$(p_1 + m_1) l_1 = (p_2 + M) l_2 \tag{16}$$

and

$$(p_1 + M) l_1 = (p_2 + m_2) l_2. \tag{17}$$

If the pointer swings near the middle of the scale with no load on the pans, we have also $p_1 l_1 = p_2 l_2$, so that (16) and (17) become

$$m_1 l_1 = M l_2$$

and

$$M l_1 = m_2 l_2.$$

Whence,

$$M = \sqrt{m_1 m_2}. \tag{18}$$

In case of a balance in ordinary adjustment m_2 will so nearly equal m_1 that we may use approximation (7), p. 7, and in place of (18) write

$$M = \tfrac{1}{2} (m_1 + m_2). \tag{19}$$

20. Precautions in the Use of a Balance. —

1. Do not place on the pans anything wet, any mercury, nor anything that might injure the pans.

2. Never change the masses on the pans nor move the rider when the beam is free to swing.

3. Never touch any standard masses with the fingers — use forceps.

4. Keep all standard masses in the proper compartments in the box when not actually in use upon the balance pan.

5. Never raise nor lower the arrestment so quickly as to cause any jerk.

6. When not actually altering masses keep the case closed.

7. Before leaving the balance bring the arrestment into play so that the beam is not free to swing, set the rider at the zero mark, dust off the pans and the floor of the case with a camel's-hair brush, and close the case.

21. Density and Specific Gravity. — If a body of mass m occupies a volume v, then the average *density* of the body is given by

$$D = \frac{m}{v}. \tag{20}$$

From this expression it is seen that the number which expresses a density depends upon the units in terms of which the mass and volume are measured. For example, at 4° C. the density of lead is about 708 pounds per cubic foot, or 2868 grains per cubic inch, or 11.34 grams per cubic centimeter. Since density is a concrete quantity, the units in terms of which the mass and volume of the body are measured must always be stated. Since most bodies change their volume somewhat with changes of temperature, the density of a substance depends upon its temperature; and so in accurate work the temperature at which a determination is made should always be stated.

The *specific gravity* or *relative density* of a substance is the ratio of its density to the density of some standard substance. In other words, the specific gravity of a body is the ratio of its mass to that of an equal volume of a standard substance. Specific gravity is thus a numerical ratio, an abstract number which is independent of the units employed. For solids and liquids, water at the temperature of its maximum density (4° C. or 39° F.) is arbitrarily taken as the standard substance.

Since in the c.g.s. system of units the unit of mass is the mass of a unit volume of water at the temperature of its maximum density, it follows that the density of a body in grams per cubic centimeter is numerically equal to its specific gravity.

22. Friction. — If a body resting upon another be acted upon by a force parallel to the surface separating them, the body will not start to move until this force has reached a certain definite value. Moreover, the force F_p which is necessary to start the body is directly proportional to the force F_n which presses the two surfaces together. That is, $F_p = \mu F_n$, in which the constant μ is called the *coefficient of static friction*. When the body does start to move, the force which is required to keep it moving uniformly is somewhat less than the force that is needed to start it. And this force F_p' which is necessary to keep the body moving uniformly is also directly proportional to the force F_n which presses the two surfaces together. That is, $F_p' = bF_n$, in which the constant b is called the *coefficient of kinetic friction*. Since F_p is greater than F_p', μ is greater than b.

23. Moment of Inertia. — If a torque be applied to a body, there will be produced an angular acceleration of its motion proportional to the torque applied. That is,

$$a = \frac{L}{K}, \tag{21}$$

where L is the torque, *i.e.*, the moment of the applied force about the axis of rotation, and K is a function of the mass and distribution of the particles composing the body called the " moment of inertia " of the body with respect to the axis of rotation. The *moment of inertia* of a rigid body with respect to a given axis is that property of the body which requires a torque to change the angular velocity of the body.

The moment of inertia of a body can be shown to be numerically equal to the sum of the products of the masses of the particles composing the body and the squares of their respective distances from the axis of rotation, *i.e.*,

$$K = \sum mr^2. \tag{22}$$

The moment of inertia of a body of simple geometric form can be computed, but the moment of inertia of an irregularly shaped body may often be determined most easily by experiment. The experimental determination is usually made by comparison with a body whose moment of inertia can be computed. For such comparisons cylinders and rings of known dimensions are convenient.

The moment of inertia of a uniform right solid cylinder of mass M and diameter d, about its geometric axis is

$$\tfrac{1}{8} M d^2. \tag{23}$$

About any axis parallel to the geometric axis and distant p from it, the moment of inertia is

$$\tfrac{1}{8} M d^2 + M p^2. \tag{24}$$

If the cylinder have a length l, the moment of inertia about an axis through the center and normal to the length is

$$M\left[\frac{d^2}{16} + \frac{l^2}{12}\right], \tag{25}$$

while about an axis coinciding with the diameter of one end, the moment of inertia of the cylinder is

$$M\left[\frac{d^2}{16} + \frac{l^2}{3}\right]. \tag{26}$$

The moment of inertia of a ring or right hollow circular cylinder of outer diameter d_o and inner diameter d_i, with respect to the geometric axis, is

$$\tfrac{1}{8} M (d_o^2 + d_i^2). \tag{27}$$

24. Elasticity. — When a body is perfectly elastic, a given deforming force keeps it distorted to the same extent no matter for how long a time the force is applied. This means that the distortion calls into play a restoring force which, so long as the body is at rest, is exactly equal and opposite to the deforming force. It follows that, when the deforming force is removed, this restoring force causes the body completely to recover its original shape and size. When a body is imperfectly elastic, a given deforming force produces a gradual yielding so that the restoring force which the distortion calls into play is in this case not quite equal to the deforming force. It follows that when the deforming force is removed from a body which is imperfectly elastic, the body does not completely recover its original shape and size. It is said to have received a *permanent set*, or to have been deformed beyond its *elastic limit*. So long as any body is not deformed beyond its elastic limit it is perfectly elastic.

The ratio of a force to the area on which it acts is called a *stress*. The ratio of a deformation to the original value of the length, volume, or whatever has been deformed, is called a *strain*. When a body has not passed its elastic limit, the ratio of the restoring stress to the strain which produced it is constant and is called a *coefficient of elasticity*.

If a wire is stretched or a pillar shortened by a load applied to it, the strain is the change of length divided by the original length. In this case the ratio of the stress to the strain is called the *tensile coefficient of elasticity* or *Young's modulus*. That is,

$$\text{Young's modulus} = \frac{\dfrac{\text{force of restitution}}{\text{area of cross section}}}{\dfrac{\text{elongation}}{\text{original length}}} \tag{28}$$

If a toy balloon were fastened under water and then pressure applied to the water, the balloon would decrease in volume without changing its shape. In this case the strain is the change in volume divided by the original volume, and the corresponding coefficient of elasticity is called the *bulk modulus*.

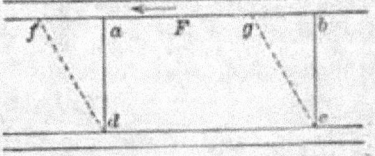

FIG. 17.

If a rectangular parallelopiped of rubber *ac* (Fig. 17) has two opposite faces glued to two boards, and if one of these boards is pushed sideways in its own plane, there is no change in the volume of the block but its shape is changed to *fgcd*. In this case the strain is the ratio of *af* to *ad*, and is called a *shear* or a *shearing strain*. If *F* is the force applied, and *A* the area of the face *ab*, then F divided by A is called a *shearing stress*. If the block of rubber is very thin in a direction normal to the paper, and if it is bent around until *ad* coincides with *bc*, it is seen that a shear is the kind of strain involved in the twisting of a wire about its geometric axis. The ratio of a shearing stress to the shearing strain which it produces is called the *simple rigidity* or the *slide modulus* of the material sheared.

25. Viscosity. — When a fluid flows along a solid surface, the layer of fluid adjacent to the solid surface is, on account of cohesion, at rest, and the speed of the other layers is greater, the greater their distance from the solid. As a result of this difference in the speed of the successive layers, the more slowly moving layer tends to retard the more quickly moving layer and is itself accelerated by the action of this layer. The property of a material fluid by which time is required for friction to produce a change in the relative motion of its parts is called *viscosity*.

In an elastic solid a shearing stress produces a shearing strain, and this strain, in turn, produces a restoring stress. If the body be subject to a given stress that is not beyond its elastic limit, the strain does not change with the lapse of time, and the ratio of the stress to the strain is a coefficient of elasticity.

In a fluid a shearing stress produces a shearing strain, and with the strain there is developed a stress that opposes the distortion but does not tend to restore the fluid to any former shape. In fact, any shearing stress, however slight, produces a continuously increasing strain, and the ratio of the shearing stress to the shearing strain thereby developed per unit time is called the *coefficient of viscosity* of the fluid.

Consider two parallel layers of fluid x cm. apart, one moving with a speed s_1 and the other with a smaller speed s_2 in the same direction. Since there is no abrupt change of velocity in passing from one layer to the other, the speed of the intervening layers varies uniformly between these values, and the change of speed per unit distance, normal to the direction of motion, equals $(s_1' - s_2)/x$. This quantity is called the "velocity gradient," and is denoted by the symbol s'.

Now the fluid on one side of a specified layer moves more

rapidly than the layer, while the fluid on the other side moves more slowly. Thus, the more rapidly moving fluid is acted upon by a retarding force, and the more slowly moving fluid by an accelerating force. If either of these forces be denoted by F and the area of the slipping surface by A, then the shearing stress is F/A (Art. 24), and the strain produced in unit time is $(s_1 - s_2)/x$. Consequently the coefficient of viscosity of the fluid is

$$\eta = \frac{F}{A} \div \frac{s_1 - s_2}{x} = \frac{F}{As'}. \tag{29}$$

That is, the coefficient of viscosity equals the tangential force per unit area per unit velocity gradient.

Exp. 1. Study of the Vernier

THEORY OF THE EXPERIMENT. — Read Art. 13.

MANIPULATION. — For each of several assigned instruments note the value in centimeters, inches, or degrees, as the case may be, of the smallest divisions on the main scale. Bring the zero line of the vernier into coincidence with a line of the main scale and note the number of divisions on the main scale that are included between the zero line of the vernier and the last line. From this result find the value of a vernier division in terms of a division on the main scale, and then, the least count of the vernier by (3) or (4).

Set the vernier at random and make a sketch of the vernier and scale as in Fig. 5 or 6. Designate the value of the scale division on either side of the zero of the vernier, and also the vernier division that coincides with a division on the main scale. Below the sketch record (a), the number of divisions on the vernier, (b), the least count of the vernier, (c), the vernier reading of the instrument as arranged in the sketch.

Exp. 2. Study of the Sensitivity of a Balance

THEORY OF THE EXPERIMENT. — Read Arts. 16 and 18. Fig. 18 represents a balance beam consisting of the frame that carries the three knife edges together with the pointer and bob. The pointer is perpendicular to the line joining the end knife edges. The distance l from the fulcrum to either of the end knife edges is called an *arm*. We shall limit our consideration to a balance with equal arms. Call the angle between the pointer and either arm, θ; the distance from the center of mass of the beam to the fulcrum, r; the mass of each pan, p; the load on the left pan, M_1;

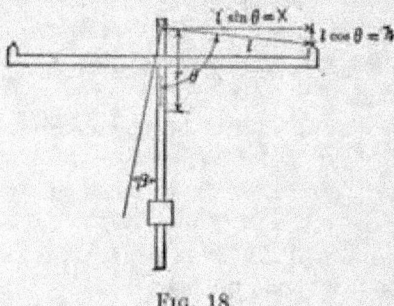

FIG. 18.

the load in the right pan, M_2; the mass of the beam, M_3; and represent the angular deflection produced by a difference of load $M_1 - M_2$ by β.

Since the distance h from the fulcrum to the line joining the end knife edge equals $l \cos \theta$, and the perpendicular distance x from one of the end knife edges to the pointer equals $l \sin \theta$, the sensitivity of an equal arm balance having pans of equal mass is, by (14),

$$\frac{\beta}{M_1 - M_2} = \frac{x}{(M_1 + M_2 + 2\,p)\,h + M_3 r}. \qquad (30)$$

It should be noted that when θ is less than 90°, h ($= l \cos \theta$) is positive; when θ is 90°, $h = 0$; when θ is more than 90°, h is negative.

An inspection of the above equation shows that with the center of mass of the beam fixed with respect to the beam,

(a), if the end knife edges are below the fulcrum, the sensitivity decreases when the mass of the pans, of the loads on the pans, or of the beam, is increased;

(b), if the three knife edges are in line, the sensitivity is independent of the masses of pans or loads on them;

(c), if the end knife edges are above the fulcrum, the sensitivity increases when the load increases.

Also, for a given load on the pans, the sensitivity may be increased by:

(d), increasing the perpendicular distance from the end knife edge to the pointer;

(e), lowering the central knife edge relative to the line joining the end knife edges;

(f), decreasing the distance r between the center of mass and the fulcrum;

(g), decreasing the mass of the beam M_2 or the mass p of the pans.

The object of this experiment is to measure the sensitivity of a model balance and also compute it by means of the above equation, (a), when the position of the center of mass of the beam is changed, the other factors remaining constant; (b), when its loads in the pans are changed while the fulcrum is below, in line with, and above, the end knife edges.

MANIPULATION. — The model balance used in this experiment is shown in Fig. 19. The beam consists of a meter bar provided with collars X and Y, each carrying three knife edges a_1, b_1, c_1, a_2, b_2, and c_2. The fulcrum K_3 is in line with knife edges b_1 and b_2. By hanging the pans on the proper knife edges, the fulcrum may be below, in line with, or above the end knife edges.

For the purpose of finding the center of mass of the beam, the plate which supports the fulcrum K_3 is provided with a knife edge K_4 on which the beam can be balanced. Since the beam is symmetrical about the pointer, and the bob B is of greater mass than the remaining parts, the center of mass lies in the axis of the pointer. By moving the bob along the pointer the center of mass can be made to coincide with the center of the groove P, Q, or R.

See that the fulcrum is midway between the two end knife edges. With the pans removed, place the groove P on the knife edge K_4 and adjust the position of the bob until the beam balances.

The center of mass is now at P.   *In balancing the beam on K_4, care must be taken that the beam does not fall.*

Replace the beam on the fulcrum K_3, and suspend the pans from a_1 and a_2.   With the pans empty, place 10 gm. on one pan and note the point of rest.   Place the 10 gm. on the other pan and

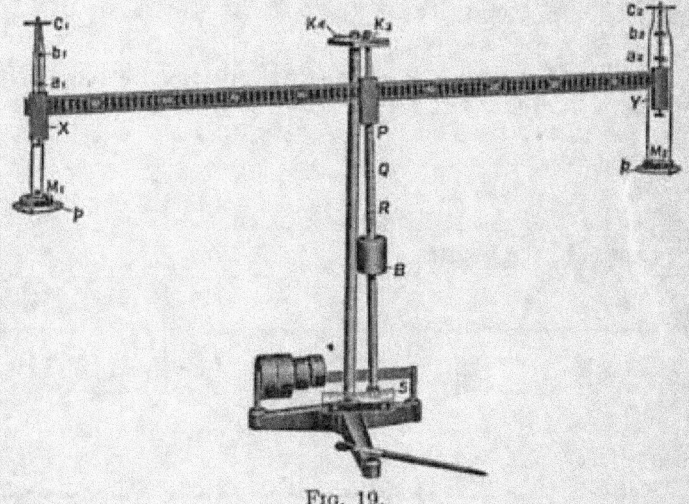

FIG. 19.

note the new point of rest.   Half the difference between these two readings, divided by 10, equals the sensitivity for zero load, expressed in radians per gram.   In the same manner, find the sensitivity for loads of 200 gm., and for 500 gm.

Suspend the scale pans from b_1 and b_2, and find the sensitivity for loads of 0, 200, and 500 gm.

Suspend the scale pans from c_1 and c_2, and find the sensitivity for loads of 0, 200, and 500 gm.

Now clamp the bob at the place that causes the center of mass of the beam to be at Q, suspend the pans from b_1 and b_2, and find the sensitivity for a load of 200 gm.

Finally, clamp the bob at the place that causes the center of mass of the beam to be at R, suspend the pans from b_1 and b_2, and find the sensitivity for a load of 200 gm.

With another balance find the mass M_1 of the beam, and the mass p of each pan. Measure the distance x from the fulcrum to the axis of the rods supporting the end knife edges, the distance h between adjacent end knife edges, and the distance r from the fulcrum to the centers of mass P, Q, and R. Substituting these values in (30), compute the sensitivity for each case that has been experimentally found. These sensitivities will be expressed in radians per gram.

The data of the experiment may be conveniently tabulated as follows:

$p = \ldots$ gm. $x = \ldots$ cm.

$M_1 = \ldots$ gm. $h = \ldots$ cm.

Length of one scale division $= \ldots$ cm. $r_P = \ldots$ cm.

Length of the pointer $= \ldots$ cm. $r_Q = \ldots$ cm.

                                                     $r_R = \ldots$ cm.

Position of end knife edges	r	$M_1 = M_2$	Point of rest	$M_1 - M_2$	Point of rest	Deflection	Sensitivity	
							Obs.	Calc.
Below fulcrum	.. cm.	.. gm.	....	10 gm.	....	....	....	....

Exp. 3. Determination of the Mass of a Body from a Weighing Reduced to Vacuo

THEORY OF THE EXPERIMENT. — Read Arts. 16, 17, 19, 20, and 21. Masses are usually compared by weighing on an equal arm balance with standard masses made of brass. Two masses on the pans of an equal arm balance will have the same apparent weight when the resultant downward force on one is equal to that on the other. Since this resultant downward force is the difference between the weight of the body and the buoyant force of the air acting upward, and since from Archimedes' Principle, the buoyant force equals the weight of air displaced by the body, it follows that two bodies of the same mass will balance one another in air only if they be of the same volume. A weighing made in air, therefore, is subject to an error due to the buoyancy of the air. This error is quite appreciable when the density of the body differs much from the density of the standard masses employed. For example, in weighing a quantity of water with standard masses made of brass, the error due to buoyancy of air amounts to about 1.06 milligram for every gram, or to about one-tenth of one per cent.

The object of this experiment is to weigh in air two bodies of different but known densities, and compute the weights that would have been obtained if the weighings had been performed in vacuo.

MANIPULATION. — Weigh each body by the Method of Vibrations, first on one pan of the balance, and then on the other. Record all observations, including the turning points of the pointer. Compute the apparent mass m of each body by means of (19), using sufficient figures to express the result to tenths of a milligram. Knowing the density D of the body under test, that of air ρ, and that of the brass standard masses d, together with the true mass m of the standards when the balance is in equilibrium, the true mass of the object M may be computed from (15).

In making the calculations above indicated, the precision of the result will depend largely upon the number of figures used to express the ratios of the densities in (15).

Problem. — Compute the true mass of a piece of cork ($D = 0.24$ gm. per cc.) which is balanced in air ($\rho = 0.0012$ gm. per cc.) by standard masses amounting to $m = 50.8654$ gm. made of brass of density $d = 8.47$ gm. per cc.

Solution. — Substituting these data in (15)

$$M\left(1 - \frac{0.0012}{0.24}\right) = 50.8654 - \frac{0.0012 \times 50.8654}{8.47},$$
$$M\ (1 - 0.0050) = 50.8654 - 0.0072,$$
$$M = 51.1138\ \text{gm.}$$

Exp. 4. Determination of the Density of a Solid by Measurement and Weighing

THEORY OF THE EXPERIMENT. — Read Arts. 11, 13, 17, 20, 21. From (20) it will be seen that the density of any solid could readily be determined if a specimen of it could be obtained in a shape such that its volume could easily be computed.

MANIPULATION. — The specimen to be used is a cylinder. Measure its diameter with a micrometer caliper and its length with a vernier caliper and calculate the volume. Determine the mass by weighing, using the method of vibrations. In order to get a very accurate value for the density it would be necessary to correct the weighing by allowing for the buoyancy of the air in the manner described in Exp. 3.

First, without making this correction, divide the apparent mass of the specimen by the volume. This gives an approximate value for the density. If a more precise value is required, use this approximate value in (15) to get the corrected mass of the cylinder, and then calculate the density by (20).

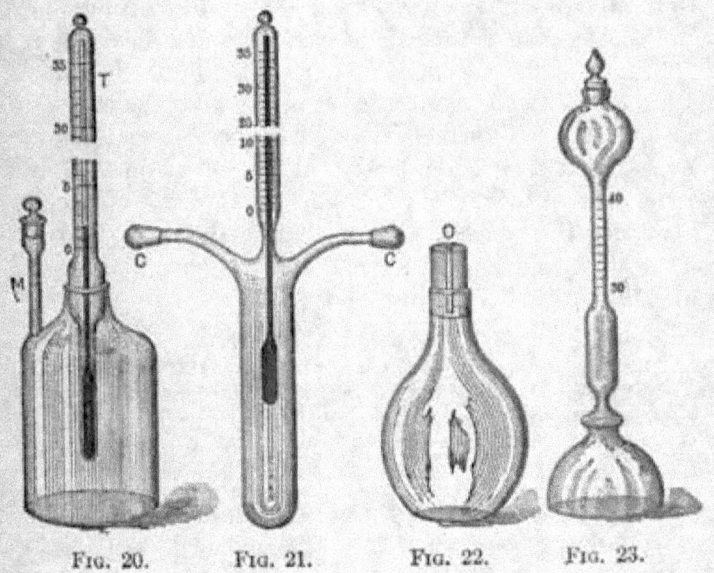

FIG. 20.   FIG. 21.   FIG. 22.   FIG. 23.

Exp. 5. Determination of the Density and Specific Gravity of a Liquid with a Pyknometer

THEORY OF THE EXPERIMENT. — Read Arts. 16, 17, 20, and 21. The pyknometer is essentially a small glass vessel of definite volume. Various forms suitable for determining the densities of liquid are given in Figs. 20–24.

The pyknometers in Figs. 21 and 24 can be used only for liquids, while the others can be used for either liquids or solids. The most common form, that shown in Fig. 22, consists of a small bottle fitted with a perforated glass stopper that always comes accurately to a seat at the same point, so that the volume of the bottle is definite when the stopper is in place. This form is often called a specific gravity bottle.

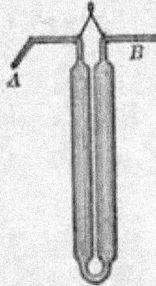

The volume of the pyknometer is obtained from two weighings, first when empty, and second when filled with a liquid of known density, e.g., water. If the mass of water contained in the filled pyknometer is denoted by W and its density by ρ_w, then the volume is

$$v = \frac{W}{\rho_w}.$$

Fig. 24.

Now let the water be replaced by the specimen. If the mass of this second liquid filling the pyknometer be denoted by s and its density by ρ_s, then

$$\rho_s \left[= \frac{s}{v} \right] = \frac{s \rho_w}{W}. \tag{31}$$

Denoting the maximum density of water by δ, we have for the specific gravity of the specimen,

$$\text{Sp. Gr.} \left[= \frac{\rho_s}{\delta} \right] = \frac{s \rho_w}{W \delta}. \tag{32}$$

In the preceding equations no account has been taken of the buoyant effect of the atmosphere on the liquids being weighed nor on the standard masses used in the weighing. In precise determinations this source of error cannot be neglected. The true weight of an object equals its apparent weight plus the weight of air displaced by it. And when the balance is in equilibrium, the apparent weight of the body equals the apparent weight of the standard masses. So that when the specimen is weighed in air, its true weight minus its loss of weight due to the buoyancy of the air equals the true weight of the standard masses minus their loss in weight. If the density of air be denoted by ρ_a and the density of the standard masses by ρ_b, this last statement says that when the pyknometer was filled with the first liquid,

$$v\rho_w - v\rho_s = W - \frac{W\rho_a}{\rho_b},$$

and when the pyknometer was filled with the second liquid,

$$v\rho_s - v\rho_a = s - \frac{s\rho_a}{\rho_b}.$$

On eliminating v from these equations, we obtain

$$\rho_s = \frac{s\,(\rho_w - \rho_a)}{W} + \rho_a.$$

MANIPULATION. — Weighing by the method of vibrations, determine first the mass p of the empty pyknometer; second, the mass $(p + W)$ of the pyknometer filled with recently boiled distilled water; and, third, the mass $(p + s)$ of the pyknometer filled with the liquid in question. Take the values of ρ_w and ρ_a from tables.

Each time before filling the pyknometer, clean it by rinsing successively with nitric acid, distilled water, and alcohol, and then dry it by putting into it the end of a tube connected to an exhaust pump. Be sure that there are no air bubbles in the pyknometer, that the outside is dry, that the stopper is in place, and that the liquid fills the capillary tube in the stopper. In order to avoid changes in volume due to changes in temperature, avoid touching the filled bottle with the bare hand.

Exp. 6. Determination of the Density and Specific Gravity of a Solid with a Pyknometer

THEORY OF THE EXPERIMENT. — Read Arts. 16, 17, 20, and 21. The object of this experiment is to determine the density and the specific gravity of a solid in small pieces.

Three suitable forms of pyknometer have already been illustrated (Figs. 20, 22, 23). To determine the volume of a solid by means of a pyknometer, four weighings are made: first, when the pyknometer is empty; second, after the specimen has been introduced; third, after the rest of the space in the pyknometer has been filled with water or other liquid; and, fourth, after the pyknometer has been emptied and then filled with the same liquid used in the third weighing.

Representing the mass of the pyknometer by the symbol p, that of the specimen by s, that of the liquid put into the pyknometer with the specimen by l, and that of the same liquid required to fill the empty pyknometer by L, we have for the four weighings:

> 1st weighing .... p,
> 2d weighing .... $p + s$,
> 3d weighing .... $p + s + l$,
> 4th weighing .... $p + L$.

From these four weighings we can find the mass s of the specimen and the mass $L - l$ of the liquid displaced by the specimen. Then, the density of the specimen

$$\rho_s \left[= \frac{\text{mass of specimen}}{\text{volume of specimen}} \right] = \frac{s\rho}{L - l}, \tag{33}$$

where ρ represents the density of the liquid.

If δ denotes the maximum density of water, it follows that the specific gravity of the specimen is

$$\text{Sp. Gr.} \left[= \frac{\rho_s}{\delta} \right] = \frac{s\rho}{(L - l)\delta}. \tag{34}$$

MANIPULATION. — Use recently boiled distilled water whenever the specimen under investigation will not be affected by water. Make all weighings by the method of vibrations. Observe all the precautions suggested in the last paragraph under Exp. 5.

Exp. 7. Determination of the Density and Specific Gravity of a Solid by Immersion

THEORY OF THE EXPERIMENT. — Read Arts. 16, 17, 20, and 21. The object of this experiment is to determine the density and specific gravity of a solid of irregular form.

Since by Archimedes' principle, a solid body immersed in a liquid is acted upon by an upthrust equal to the weight of the liquid displaced by the body, it follows that if this upthrust is measured, the weight of the displaced liquid is known, and if the weight of a unit volume of the liquid is also known, then the volume of the liquid displaced — and, therefore, the volume of the body — can be calculated. If the weight of the body in air is denoted by B_a, and its weight when immersed in the liquid by B_l, then the up-thrust of the liquid — and, consequently, the weight of the liquid displaced — is $B_a - B_l$. So that, if w denotes the weight of a unit volume of the liquid at the temperature of the experiment, the volume of the liquid displaced — and, consequently, the volume of the body — is

$$v = \frac{B_a - B_l}{w}.$$ (35)

It follows, if s denotes the mass of the specimen, that the density of the specimen is

$$\rho_s \left[= \frac{s}{v} \right] = \frac{sw}{B_a - B_l} = \frac{B_a \rho_l}{B_a - B_l},$$ (36)

the last equation in (36) being true because $s = B_a/g$ and $w = \rho_l g$.

Since specific gravity is defined as the ratio of the density of the substance in question to the maximum density, δ, of water,

$$\text{Sp. Gr.} \left[= \frac{\rho_s}{\delta} \right] = \frac{B_a \rho_l}{(B_a - B_l)\delta}.$$ (37)

When the body is lighter than the liquid in which it is to be immersed, a sinker is attached. Weighings are made to determine: first, the weight of the body in air, B_a; second, the weight of the sinker immersed in the liquid, S_l; and third, the weight of the two together when immersed, $(B + S)_l$. The weight of the body alone when immersed in the liquid is negative, but its value, sign included, is

$$B_l = (B + S)_l - S_l,$$

and this value can be substituted in (36) and (37), giving

$$\rho_s = \frac{B_a \rho_l}{B_a - (B + S)_l + S_l}$$ (38)

and $$\text{Sp. Gr.} \left[= \frac{\rho_s}{\delta} \right] = \frac{B_a \rho_l}{[B_a - (B + S)_l + S_l]\delta}.$$ (39)

MANIPULATION. — The liquid in which the body is immersed must be one which will not dissolve the body, act upon it chemically, nor cause it to change its volume. Whenever possible, use is made of water which has been freed of dissolved gases by boiling. If the liquid contains dissolved gases, bubbles will collect on the immersed body, causing an increased upward thrust, and therefore an error in the result. Water should be boiled for about half an hour and then cooled to the temperature at which the experiment is performed. As water slowly dissolves air, it must be boiled on the day it is used.

The motion of the balance beam is so much damped by the immersion of the load in a liquid that it is useless to weigh by the method of vibrations. The values of ρ_l and δ are to be taken from tables.

Exp. 8. Determination of the Specific Gravity of a Liquid with the Mohr-Westphal Balance

THEORY OF EXPERIMENT. — Read Art. 21. The object of this experiment is to determine the specific gravity of an aqueous solution by means of a Mohr-Westphal balance.

From Archimedes' principle it follows that if a body of constant volume be immersed in various liquids, the corresponding losses of weight sustained by the body will represent the weights of equal volumes of the various liquids. Whence, if a body of volume v, when immersed in succession in two liquids, of densities ρ_1 and ρ_2, sustain the respective losses of weight w_1 and w_2, then

$$\frac{w_1}{w_2} = \frac{v\rho_1 g}{v\rho_2 g} = \frac{\rho_1}{\rho_2}. \tag{40}$$

If the second liquid be water at the temperature of its maximum density, then the ratio of w_1 to w_2 gives the specific gravity of the first liquid.

If, therefore, a means be devised for measuring the loss of weight of a given body when immersed in any liquid, and also for determining what loss the same body would suffer if it were immersed in water at 4° C., the specific gravity of the liquid could be computed by means of the above equation.

A convenient instrument designed for the purpose is the Mohr-Westphal balance. This device (Fig. 25) consists of a decimally divided balance beam at one end of which is suspended a glass sinker for immersion. The other end of the beam is so counter-balanced that the beam is held in equilibrium when the sinker is surrounded by air. The instrument is also provided with four riders which are ordi-narily equal in mass to 1.0, 0.1, 0.01, and 0.001 *of the mass of water dis-placed by the sinker.* Thus, if the sinker be immersed in water, one unit rider placed at the end of the beam would be required to compen-sate for the loss sustained by the sinker and to bring the beam back to a horizontal position. Again, if with the sinker immersed in a certain liquid the beam is brought into a horizontal position when a unit rider is hung on the hook *A*, the tenths rider on the

Fig. 25.

second notch *C*, and the hundredths rider on the third notch *B*, the theory of moments of forces shows that the upthrust on the sinker is 1.023 times as great as in the preceding case. Consequently the specific gravity of the given liquid is 1.023.

If, as the temperature rises, the sinker were to expand at the same rate that water does, the temperature at which the Mohr-Westphal balance is used would make no difference, for the sinker would always displace the same mass of water. But, as a matter of fact, at ordinary room temperatures water expands more rapidly than glass, so that when the temperature is a little above

20° C. the Mohr-Westphal balance reads 0.1 per cent lower than it would at 15°. Moreover, the temperature in a laboratory is usually not so low as 4° C., and so the riders are usually adjusted to read specific gravities with reference to water at 15° — about the temperature at which European laboratories are usually kept. In order to use the balance in a laboratory at about 20° and to get specific gravities with reference to water at 4° it will then be necessary to apply a correction.

To find what this correction is, let b_{15} and b_t denote the respective readings of the balance when the sinker is immersed, (a) in water of density ρ_{15} at 15°, and (b) in the liquid whose density ρ_t at $t°$ is desired, and let v_{15} and v_t denote the respective volumes of the sinker. Then the weights of liquid displaced by the sinker in the two cases are respectively $\rho_{15}v_{15}g$ and $\rho_t v_t g$. Since the readings of the balance are proportional to these weights,

$$Kb_{15} = \rho_{15}v_{15}g = \rho_{15}v_0 g\,(1 + \gamma \cdot 15) \tag{41}$$

and
$$Kb_t = \rho_t v_t g = \rho_t v_0 g\,(1 + \gamma t), \tag{42}$$

where v_0 denotes the volume of the sinker at 0° and γ its coefficient of expansion. On dividing each member of (42) by the corresponding member of (41), we obtain

$$\frac{b_t}{b_{15}} = \frac{\rho_t\,(1 + \gamma t)}{\rho_{15}\,(1 + \gamma \cdot 15)}.$$

Whence, since the balance is so adjusted that $b_{15} = 1$,

$$\rho_t = \rho_{15}b_t\left(\frac{1 + \gamma \cdot 15}{1 + \gamma t}\right),$$

or, employing approximation (5), p. 7,

$$\rho_t \doteqdot \rho_{15}b_t\,(1 + \gamma \cdot 15)\,(1 - \gamma t),$$

or, employing approximation (2), p. 7,

$$\rho_t \doteqdot \rho_{15}b_t\,[1 - \gamma\,(t - 15)]. \tag{43}$$

If the specific gravity of the liquid is desired, we have at once, if δ denotes the maximum density of water,

$$\text{Sp. Gr.} \left[= \frac{\rho_t}{\delta}\right] = \frac{\rho_{15}b_t}{\delta}\,[1 - \gamma\,(t - 15)]. \tag{44}$$

Since γ is small and ρ_{15} differs only slightly from δ, it will be seen that if only fairly accurate values are desired, (43) and (44) give

$$\rho_t \doteq \delta b_t \tag{45}$$

and
$$\text{Sp. Gr.} \doteq b_t. \tag{46}$$

MANIPULATION. — With the sinker in air and no rider on the beam, the instrument is first leveled until the pointer attached to the beam indicates zero. The sinker is then immersed in the liquid whose specific gravity is to be determined, and riders are placed in the notches on the beam until the pointer again indicates zero.

Exp. 9. Calibration of an Hydrometer of Variable Immersion

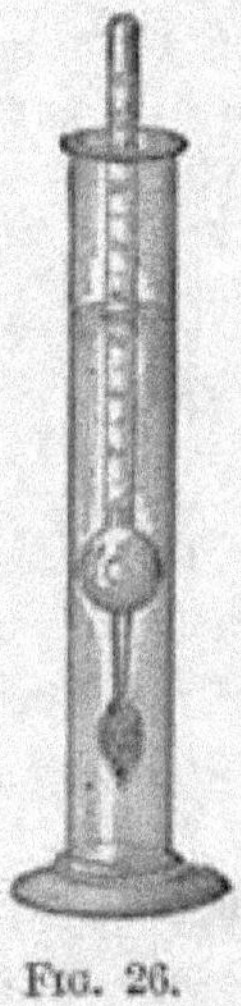

THEORY OF THE EXPERIMENT. — Read Art. 21. In the measurement of the specific gravity of liquids for technical purposes where great accuracy is unnecessary, some form of hydrometer of variable immersion is usually employed. The hydrometer (Fig. 26) consists of a closed graduated glass tube of uniform cross section with a weighted bulb on the lower end. The mass and volume of the instrument are so chosen that when it is placed in the liquid whose specific gravity is to be determined it will float upright. The specific gravity of the liquid is

FIG. 26.

shown by the depth to which the hydrometer sinks. If the graduations on the stem are so spaced and numbered as to give directly the density of the liquid, the instrument is called a densimeter. Often, however, the graduations are equidistant and are referred to some arbitrary scale. Thus we have the scales of Baumé, Beck, Cartier, and Twaddell. The specific gravities corresponding to readings on these various scales are given in Table 6. Not infrequently the stem of the hydrometer contains two or more scales. When graduated with especial reference to use with some particular class of liquids, the hydrometer is called the alcoholimeter, salinimeter, etc.

The quantity which has to be added to a reading in order to obtain the corrected reading is called the *correction* for that reading. The object of this experiment is to plot a correction curve, coördinating the hydrometer readings and the corrections to be applied.

(a) *Scale with divisions of equal length.* If an hydrometer of mass m sinks to scale division d_1 when placed in a liquid of density ρ_1, and to division d_2 when placed in a liquid of density ρ_2, then by Archimedes' principle the volume of the first liquid displaced is $\dfrac{m}{\rho_1}$ and of the second is $\dfrac{m}{\rho_2}$. If u denotes the volume of that part of the stem which is included between two consecutive scale divisions, then

$$\frac{m}{\rho_2} = \frac{m}{\rho_1} - u\,(d_1 - d_2).$$

Whence,
$$u = \frac{m\,(\rho_2 - \rho_1)}{\rho_1\rho_2\,(d_1 - d_2)}, \tag{47}$$

or
$$\rho_2 = \frac{m\rho_1}{m - \rho_1 u\,(d_1 - d_2)}. \tag{48}$$

From (47), if ρ_1, ρ_2, and m are known, the value of u can be found, and from (48), if u, ρ_1, and m are known, ρ_2 can be found.

If the maximum density of water is denoted by δ, the specific gravity of the second liquid is

$$\text{Sp. Gr.}\left[= \frac{\rho_2}{\delta}\right] = \frac{m\rho_1}{[m - \rho_1 u\,(d_1 - d_2)]\,\delta}. \tag{49}$$

(b) *Scale in which the successive divisions express equal differences in density.* Consider a wooden rod of mass m, of uniform cross section q, and so loaded at one end that it will float upright. When the rod floats, the weight of liquid displaced is by Archimedes' principle equal to the weight of the rod. That is, if the rod sinks a distance l_1 in a liquid of density ρ_1,

$$\rho_1 l_1 q g = mg.$$

Whence,
$$l_1 = \frac{m}{\rho_1 q}. \tag{50}$$

Similarly, if the rod sinks a distance l_2 in a liquid of density ρ_2,

$$l_2 = \frac{m}{\rho_2 q}.$$ (51)

Dividing each member of (50) by the corresponding member of (51),

$$\frac{l_1}{l_2} = \frac{\rho_2}{\rho_1}.$$ (52)

That is, the distances to which this hydrometer of uniform cross section sinks in various liquids are inversely proportional to the densities of those liquids.

Consider now an hydrometer of the usual form, which is not of uniform cross section throughout, but which is of uniform cross section above some point K (Fig. 27). For this hydrometer there is at some unknown distance x below K a point to which the hydrometer would extend if it had still the same mass and volume which it really has, but if, instead of the varying cross section which it really has, it continued throughout with the same cross section which it has above K. Suppose that in one liquid this hydrometer sinks to a point distant h_1 above K, and in another liquid to a point distant h_2 above K. Then from (52),

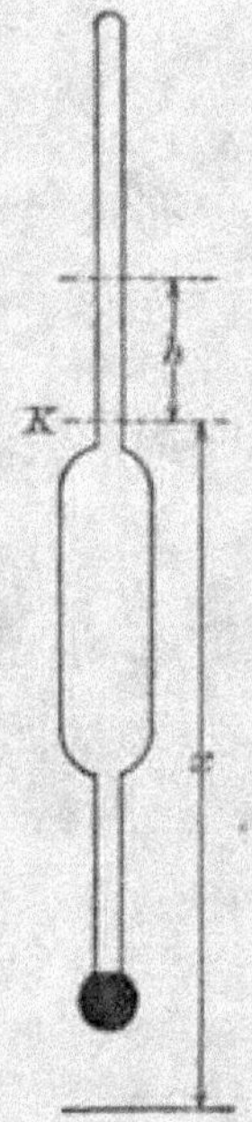

$$\left[\frac{l_1}{l_2} = \right]\frac{h_1 + x}{h_2 + x} = \frac{\rho_2}{\rho_1},$$ (53)

or

$$x = \frac{h_2\rho_2 - h_1\rho_1}{\rho_1 - \rho_2}.$$ (54)

Also, from (53)

$$\rho_2 = \rho_1 \cdot \frac{h_1 + x}{h_2 + x}.$$ (55)

If the maximum density of water be denoted by δ the specific gravity of the liquid is, then,

$$\text{Sp. Gr.}\left[= \frac{\rho_2}{\delta}\right] = \frac{\rho_1 (h_1 + x)}{\delta (h_2 + x)}.$$ (56)

Fig. 27.

That is, for any given distance h_2 above K, Fig. 27, the hydrometer reading should be that given by (56).

If we determine to what distance above K the hydrometer sinks in each of two liquids of known densities, we can by (54) determine x. And if we know to what distance above K the hydrometer sinks in one liquid of known density, and know also x, then if we determine to what distance above K the hydrometer sinks in any other liquid, we can by (56) determine the specific gravity of that liquid.

Uniformity of cross section of the hydrometer may be tested by reading diameters at various points with a micrometer caliper. If the cross section be not uniform above K, the above method of calibration is not applicable. In this case some dozen or twenty solutions having densities varying somewhat uniformly within the range of the hydrometer should be made up, the density of each determined, and the reading of the hydrometer in each taken. This method of calibration is, of course, more accurate, but is more tedious than the other.

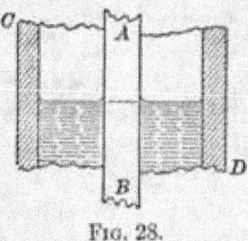

Fig. 28.

MANIPULATION. — The surface of the liquid about an hydrometer is usually of a shape similar to that in Fig. 28. AB is the stem of the hydrometer and CD is a tall narrow jar in which the liquid is placed. First be sure that the hydrometer is floating freely, and then place the eye below the level of the liquid surface and raise it until it is sighting the hydrometer along the dotted line. The point of the scale crossed by this line is the required reading. The temperature of the liquid should be noted at the time of each observation. When changing from one liquid to another, the jar, hydrometer, and thermometer are to be thoroughly washed and dried. Determine the densities of two liquids either with a pyknometer or with a Mohr-Westphal balance. Observe the scale readings on the hydrometer when it is floated in turn in the two liquids.

(a) If the hydrometer has a scale with equal divisions, weigh the instrument, place it in succession in two liquids of known densities, and then by means of (47) calculate the value of u. By means of (49) calculate the specific gravity corresponding to each of the numbered scale divisions on the stem of the hydrometer. Plot a curve with these calculated specific gravities as abscissæ and the corresponding scale readings as ordinates. This is the calibration curve of the instrument. The calibration curve should be checked by comparing two or three values obtained by means of the hydrometer in connection with the curve, with values obtained by means of a pyknometer or a Mohr-Westphal balance.

(b) In the case of the densimeter or direct-reading hydrometer, select any convenient point on the scale as K. Lay a steel scale along the stem of the hydrometer and record the distance from K to each numbered division of the hydrometer. Also, record the distance from K to each of the points to which the hydrometer sank when placed in the two liquids whose densities were previously determined. From these last two readings together with the densities already determined, calculate x by (54). Knowing x and the distance from K to the various hydrometer divisions, use (56) to determine what the hydrometer readings should be at the various points along the scale.

The observations and results should be arranged in a table, somewhat as follows:

Hydrometer reading $= H$	Distance above $K = k$	$k + z$	Specific gravity $S = \dfrac{s_1(k_1 + z)}{\delta(k + z)}$	Correction $= S - H$

Plot a correction curve, coördinating hydrometer readings and the corrections to be applied.

Exp. 10. Determination of the Correction Factor of a Planimeter

THEORY OF THE EXPERIMENT. — Read Art. 15. The correction factor of a planimeter is that number — usually near unity — by which the area read from the instrument must be multiplied in order to get the true area.

Equation (10) suggests at once a method of determining the correction factor of a planimeter. If d denotes the diameter of the roller, and l the length of the tracer arm AC, then the area which can just be circumscribed by the tracing point while the roller rotates once is, by (10), equal to πdl. If the roller be so graduated that the area indicated for one rotation is a, the correction factor K is given by

$$K = \frac{\pi dl}{a}. \tag{57}$$

MANIPULATION. — With a steel scale and a sharp pencil lay off a rectangular area of not less than 150 sq. cm. Make five careful readings of the length and breadth of the rectangle. If the tracer arm be adjustable in length, note the reading on its scale. Place the pole point outside the rectangle, bring the tracing point to one corner, and read the planimeter. Using the steel scale as a straight edge to guide the tracing point, circumscribe the rectangle in the clockwise direction, and again read the planimeter. In this manner measure the area at least ten times. The product of the average length and average breadth of the figure divided by the average difference between the final and initial readings of the planimeter gives the correction factor.

With a micrometer caliper determine the diameter of the roller. With the steel scale make five readings of the length of the tracer arm. From these calculate the correction factor by (57). Compare the results obtained by the two methods.

Exp. 11. Correction for Eccentricity in the Mounting of a Divided Circle

THEORY OF THE EXPERIMENT. — Angles are often measured by means of a divided circle and an index or vernier attached to an arm capable of rotation about an axis passing through the center of the circle. This method is subject to a source of error due to the mechanical difficulty of mounting the arm carrying the vernier so that its axis of rotation accurately coincides with the normal axis of the divided circle. The object of this experiment is to construct a correction curve for a divided circle having an eccentrically mounted vernier.

Let C (Fig. 29) be the center of the divided circle, A and B the zero points of the two verniers carried on arms capable of rotation about the point D. If the line AB passes through D, and D coincides with C, there is no eccentricity in the mounting, and correct angular readings are obtained by means of a single vernier. But in the general case where neither of these conditions is fulfilled, correct angular readings can be obtained only from simultaneous readings of the two verniers A and B.

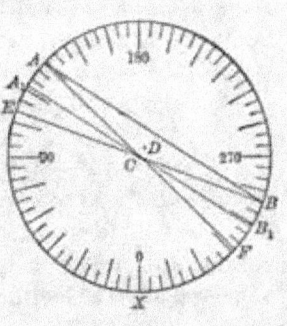

Fig. 29.

Let $A°$ and $B°$ be the observed readings. Draw A_1B_1 through C parallel to AB. If there were no eccentricity in the mounting, and if A and B were diametrically opposite, the readings would be $A_1°$ and $B_1°$. In other words, $A_1°$ and $B_1°$ are the true readings corresponding to the observed readings $A°$ and $B°$. Through C draw the lines BE and AF.

Since A_1B_1 is parallel to AB, and AC equals BC,

$$\angle ECA_1 = \angle CBA = \angle BAC = \angle ACA_1.$$

Therefore $\qquad \angle XCA_1 = \tfrac{1}{2}(\angle XCE + \angle XCA),$

or $\qquad\qquad A_1° = \tfrac{1}{2}(E° + A°).$

If the division lines on the circle are numbered as shown in the figure, $E° = B° - 180°$. Consequently the corrected reading of the vernier A is

$$A_1° = \tfrac{1}{2}(A° + B° - 180°). \qquad (58)$$

This is the corrected reading for the vernier giving the smaller reading.

In precisely the same manner, since $B_1° = \tfrac{1}{2}(B° + F°)$ and since $F° = 180° + A°$, the corrected reading of the vernier B is

$$B_1° = \tfrac{1}{2}(A° + B° + 180°). \qquad (59)$$

This is the corrected reading for the vernier giving the larger reading.

In this manner, by means of two verniers, is obtained the reading of either vernier corrected for eccentricity of mounting.

MANIPULATION. — Starting with one vernier near the zero point of the circle, read both verniers. Then move the verniers about thirty degrees and again read them both. Repeat at intervals of about thirty degrees until the entire circumference is traversed. The corrections for the observed vernier readings are found by subtracting the observed readings from the corrected readings.

On cross-section paper lay off the observed readings of one vernier on the axis of abscissæ and the corresponding corrections on the axis of ordinates. The curve drawn through the points thus obtained is the correction curve for this vernier. From the form of this curve decide whether C and D are coincident, and whether AB passes through D.

Exp. 12. Radius of Curvature and Sensitivity of a Spirit Level

THEORY OF THE EXPERIMENT. — In many measurements in which a spirit level is used in connection with other physical apparatus it is necessary that the sensitivity of the level be at least as great as that of the other apparatus. An example is the case of the telescope and level of an engineer's transit. When used in leveling or in measuring vertical angles, the least vertical motion of the telescope which can be detected by means of the cross hair should a so make itself evident by a displacement of the level bubble. A test of the suitability of a level for a particular use includes the determination of the uniformity of the run of the bubble in the vial and the sensitivity of the spirit level. The *sensitivity* of a spirit level may be defined as the distance the bubble moves for an inclination of the level of one minute. Since the sensitivity can be proven to be directly proportional to the radius of curvature of the vial, it is often designated by the radius of curvature. The object of this experiment is to make a test of a spirit level.

In the laboratory a spirit level is usually tested by means of a

Level Trier consisting of a base plate upon which rests a T-shaped casting supported by two projecting steel points E and F (Fig. 30) at the end of the arms of the T and a micrometer screw M at the foot of the T. The pitch of the micrometer screw must be measured and also the perpendicular distance from the micrometer screw to the line connecting the points E and F. The level to be tested, L, is placed on the T and the position of the bubble in the

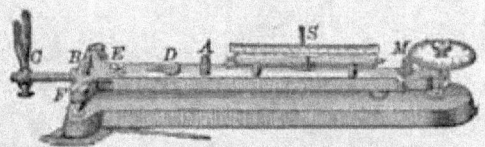

FIG. 30.

vial is noted by means of a scale engraved upon the glass or by a scale S attached to the level trier. In case it is inconvenient to separate the level from a piece of apparatus of which it forms a

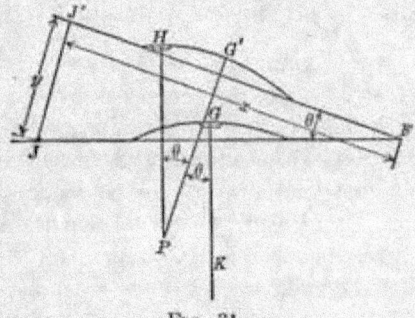

FIG. 31.

part, the entire apparatus, *e.g.*, a telescope or theodolite, may be mounted in the grooves ABC or DEF.

After the spirit level is in place, the micrometer reading is noted. The T is now tilted through a small angle by turning the micrometer screw, and readings are again taken of the micrometer screw and the position of the bubble.

Suppose that by means of the micrometer screw the T of the level trier is moved from the position FJ (Fig. 31) to the position FJ', the middle of the bubble moving meantime from G to H. If a vertical line GK were drawn through G before the micrometer screw was turned, and if this line were to move with the level, it would after the movement be in a position $G'P$, such that the angle through which it moved would equal the angle JFJ' through which the level moved. A vertical line through the middle of the bubble's position of rest has the direction of a radius of the bubble vial. If, then, HP is drawn vertically through H, both HP and $G'P$ are radii of the vial. But since HP and GK are parallel, the angle $G'PH$ equals the angle between GK and $G'P$, which latter has just been shown to equal JFJ'. It follows that

$$\angle G'PH = \angle JFJ'.$$

Let $G'P$ be denoted by R, $G'H$ by d, FJ' by x, and JJ' by y. Then

$$\frac{d}{R} = \theta \text{ radians,} \tag{60}$$

and, since the screw is always perpendicular to the T,

$$\frac{y}{x} = \tan \theta.$$

Since θ is always very small, $\tan \theta \doteqdot \theta$, and we have

$$\frac{d}{R} \doteqdot \frac{y}{x}, \tag{61}$$

whence,

$$R \doteqdot \frac{xd}{y} \doteqdot \frac{d}{\theta}. \tag{62}$$

Since, by definition, the sensitivity

$$S = \frac{d}{\theta},$$

it follows that

$$S \doteqdot R.$$

If the angle θ is to be expressed in degrees instead of in radians,

$$S\left[=\frac{d}{\theta \text{ (radians)}}\right]=\frac{d}{\dfrac{180}{\pi}\theta^{\circ}}.$$

The sensitivity of a spirit level is frequently expressed in centimeters per minute of arc. In this case,

$$S=\frac{d}{\dfrac{(180) 60}{\pi}\theta'}=\frac{d}{3438\,\theta'}=\frac{\pi d}{3438\,y}=\frac{R}{3438}. \tag{63}$$

MANIPULATION. — Place the T-shaped casting upon a piece of bristol board, and by means of slight pressure obtain an impression of the three supporting points. Measure the perpendicular distance from the impression made by the end of the micrometer screw to the line connecting the impressions of the other two supporting points.

The pitch of the micrometer screw may be obtained in the following manner: After placing the spirit level on the trier, adjust the micrometer screw until one end of the bubble is directly under a scale division near the middle of the vial; then insert under the micrometer screw a small piece of plate glass whose thickness has been already measured with a spherometer or micrometer caliper, and again adjust the micrometer screw until the bubble rests at the same point as before. The thickness of the glass plate divided by the necessary number of turns of the micrometer screw gives the pitch of the latter.

Again adjust the micrometer screw until one end of the bubble is directly under a scale division near one end of the vial. Observe the micrometer screw reading and the scale readings at both ends of the bubble; rotate the micrometer screw through a convenient number of spaces and take readings as before. Rotating the micrometer screw the same number of spaces each time, continue this operation until the bubble has been removed in some half dozen steps to the other end of its run, and then return step by step in the same manner. Repeat this series of readings three times. A series of such readings may be conveniently tabulated in the following form:

Number of observation	Micrometer reading	Readings of bubble		Displacements		Length of bubble
		Left end	Right end	Left end	Right end	
	mm.	mm.	mm.	mm.	mm.	mm.
1	3.700	1.3	10.2	. . . . . .	. . . . . .	8.9
2	3.800	6.1	14.9	4.8	4.7	8.8
3	3.900	11.1	19.8	5.0	4.9	8.7
4	4.000	16.2	25.1	5.1	5.3	8.9
5	4.100	21.1	30.2	4.9	5.1	9.1
6	4.000	16.1	25.1	5.0	5.1	9.0
7	3.900	11.1	19.9	5.0	5.2	8.8
8	3.800	6.2	15.0	4.9	4.9	8.8
9	3.700	1.3	10.2	4.9	4.8	8.9
			Mean	4.95	5.00	8.88

The values in columns 2, 3, and 4 are read, and those in 5, 6, and 7 are calculated from these readings. The values in columns 5 and 6 show the uniformity of the run of the bubble, or the variation in sensitivity when the bubble is at different positions in the vial. The average radius of curvature and sensitivity of the vial are obtained by substituting for d in (62) and (63) the mean displacement obtained from columns 5 and 6.

Care must be taken to keep the entire vial at the same temperature. It must not be touched by the fingers nor breathed upon, as when unequally heated the bubble tends to move toward the point of highest temperature.

Exp. 13. The Acceleration due to Gravity by Means of a Simple Pendulum

THEORY OF THE EXPERIMENT. — In elementary text-books on General Physics it is shown that the period of a complete to-and-fro vibration of a simple pendulum of length l vibrating through a small arc at a place where the acceleration due to gravity is g, is

$$T = 2\pi\sqrt{\frac{l}{g}}.$$

Whence,

$$g = \left(\frac{2\pi}{T}\right)^2 l. \tag{64}$$

The equation is deduced on the assumption that the pendulum has its mass concentrated at a point on the end of a perfectly flexible suspension. An increase either in the size of the bob or in the mass of the suspending wire increases the error introduced by using the above equation. If the length of the pendulum is taken as the distance from the supporting knife edge to the center of mass of the bob, and if this distance be about 100 cm., and the diameter of the bob about 3 cm., the value found for g is about 0.01 per cent too small. With the same length of pendulum, if the mass of the supporting wire be about 0.3 g. and the mass of the bob about 75 g., the value found for g is about 0.07 per cent too large.

As the period T of the simple pendulum enters (64) as a second power while the length enters as a first power, the value of the period must be known to a higher degree of precision than the value of the length l.

The method of coincidences now to be described is a very accurate method for the comparison of two nearly equal periods of vibration. In the present experiment the period of the simple pendulum is to be compared with that of a standard clock pendulum that beats seconds. If the simple pendulum swings slightly faster than the clock pendulum, a moment will occur when both are at their lowest points at the same time. But since the simple pendulum is all the time gaining on the clock pendulum, after a certain interval it will have gained a whole oscillation, and then both pendulums will again be at their lowest points. If between two such coincidences the pendulum has made n swings, then the clock pendulum has made $n - 1$ swings. One swing of the clock pendulum represents one second. Represent the time of one swing of the simple pendulum by $\frac{1}{2} T$ sec. Then the number of seconds between two coincidences is

$$\left(\frac{T}{2}\right) n = 1 \, (n - 1),$$

or
$$\frac{T}{2} = \frac{n - 1}{n} \text{ sec.} \tag{65}$$

One method of determining the instant of coincidence employs an electric circuit containing the two pendulums, a battery, and a telegraph sounder or telephone receiver, all in series as shown in Fig. 32. When the two pendulums are in coincidence, they pass through the mercury contacts A and B at the same instant, and at this instant the sounder clicks. It is to be kept in mind that

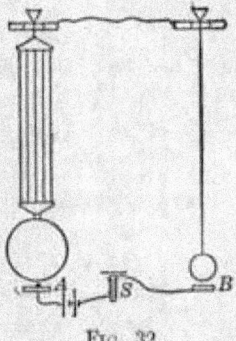

FIG. 32.

the n in the above expressions denotes the number of swings made by the simple pendulum — not by the clock pendulum.

Since one pendulum gains only slightly on the other, and since the passage of the pendulums through the mercury cups at A and B is not instantaneous, there are often clicks for several successive swings. The mean time of the first and last of these successive clicks is used as the instant of coincidence.

The actual instant of coincidence, *i.e.*, the instant when each pendulum is distant from its position of rest by the same fraction of a vibration that the other is, may occur when both pendulums are in some position other than at their lowest points, but it can never be more than half a swing from the lowest point. If there are only a few successive clicks, it will be safe to assume that in taking the mean of several successive clicks, the time of coincidence is not in error by so much as one swing. If the simple pendulum be swinging faster than the clock pendulum, the error introduced into the value for the half period by getting for n one swing too few is the difference between the period found, $\dfrac{n-2}{n-1}$, and the true period, $\dfrac{n-1}{n}$, *viz.*,

$$\frac{n-2}{n-1} - \frac{n-1}{n} = -\frac{1}{n(n-1)}.$$

If n be large compared with unity, the error is almost $-\dfrac{1}{n^2}$. Thus if $n = 70$, the error introduced into the period by an error of 1 in the number of seconds between coincidence is about -0.0002 sec. If n be small, the accuracy may be increased by counting the number of seconds to some later coincidence instead of to the second. In this case one pendulum will have gained on the other more than one swing, and the above formulas must be modified accordingly.

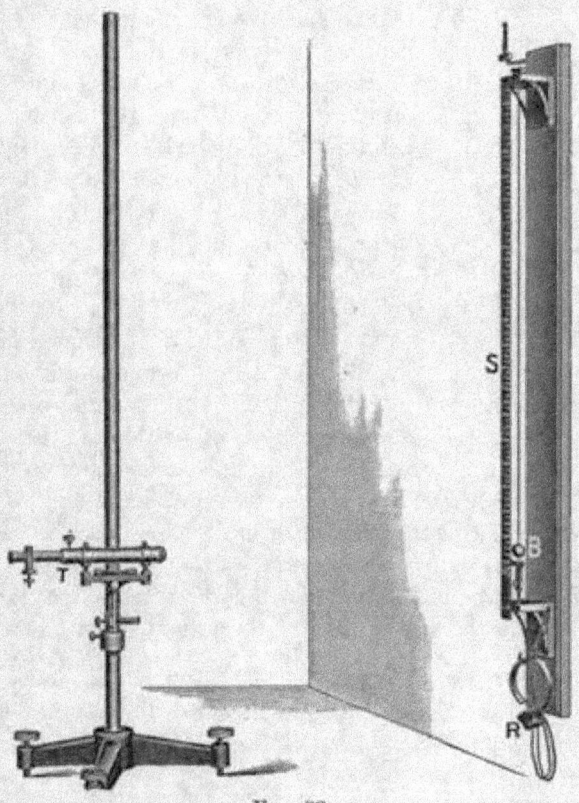

FIG. 33.

MANIPULATION. — A convenient arrangement of apparatus for the measurement of the length and the comparison of the period of a simple pendulum is shown in Fig. 33. The simple pendulum B is connected in series with the seconds pendulum of a standard clock not shown in the figure, a battery, and a telephone receiver R. Beside the simple pendulum is a vertical scale S. The length of the simple pendulum is determined by means of this scale and a horizontal telescope, T, that can be moved up and down a vertical rod.

With the pendulum at rest, adjust the position of the supporting knife edge till, (a), the knife edge is at right angles to the plane of vibration of the pendulum, (b), the contact point attached to the bob dips into the center of the ridge of mercury below it. Adjust the scale until the face of the scale and the pendulum are in the same plane, and the edge of the scale is parallel to the pendulum. Set the pendulum swinging with an amplitude of about two centimeters. By the aid of a watch held in the hand determine whether the period of the pendulum is greater or less than one second. Now, with the head band of the telephone receiver in place, note the number of swings of the simple pendulum between clicks. At each time of coincidence there will be several clicks. The mean time of the first and last of these successive clicks is to be used as the instant of coincidence.

Thus, for a given simple pendulum, indicating the swing at which the first click occurred by zero, and thereafter counting the swings for several minutes and noting the particular swings at which clicks occurred, the data given in the following table were obtained. Also, by counting the number of swings made in two minutes, it was found that the period of this pendulum was less than that of the standard clock.

Initial click of series	Final click of series	Mean	Interval
0	3	1.5	52.
52	55	53.5	52.
104	107	105.5	52.
156	159	157.5	52.
208	211	209.5	51.5
259	263	261.0	51.5
310	315	312.5	52.5
361	369	365.0	
			51.93

Since the simple pendulum had a period less than the standard clock, while the simple pendulum was making 52 vibrations the standard clock was making 51 vibrations.   Whence, the period T of the simple pendulum had the value given by (65),

$$\frac{T}{2}\left[=\frac{n-1}{n}\right]=\frac{51}{52} \text{ sec.}$$

To measure the length of the simple pendulum, first make the telescope horizontal and the supporting rod vertical in the manner described in Art. 14.   Now focalize the telescope on the supporting knife edge and adjust the position of the telescope till the cross hairs coincide with the image of the edge.   Rotate the telescope about the vertical supporting rod until the scale S is in the field of view, and take the scale reading that coincides with the cross hair.   Fractions of the smallest scale divisions are to be determined by the micrometer in the telescope eyepiece.   In the same manner read the positions of the top and of the bottom of the bob.   For the length of the simple pendulum use the distance from the knife edge to the center of the bob.   Make at least two determinations of the length and take the mean.

Exp. 14. Determination of the Speed of a Projectile by the Ballistic Pendulum

THEORY OF THE EXPERIMENT. — The object of this experiment is to determine the speed of a bullet from a rifle.

Newton proved that if two bodies are moving along the same straight line, the speed of the first with respect to the second after a collision between the two is directly proportional to the speed before the collision, the proportionality factor depending upon the elasticity of the two bodies and being called the coefficient of restitution of the given bodies.   He also proved that if no external forces act upon a system of bodies, the total momentum of the system is constant.

Imagine that a projectile of mass m and speed u strikes a body of mass M and speed U and that after the impact the speeds are u' and U' respectively.   Then before impact the speed of the projectile with respect to the other body is $(u - U)$, and after impact it is $(u' - U')$. It follows, then, from the statements in the preceding paragraph, that

$$u' - U' = e(u - U) \tag{66}$$

and
$$mu' + MU' = mu + MU, \tag{67}$$

where e is the coefficient of restitution of the bodies. If the bodies are perfectly elastic, $e = 1$, and if they are perfectly inelastic, $e = 0$. If the experiment be so arranged that the initial speed of the large mass is zero, and that after the impact the two masses move together, thus acting like inelastic bodies, then $U = 0$ and $e = 0$. On making these substitutions in (66) and (67) and then eliminating u' between them, we get

$$u = \left(\frac{m + M}{m}\right) U'. \tag{68}$$

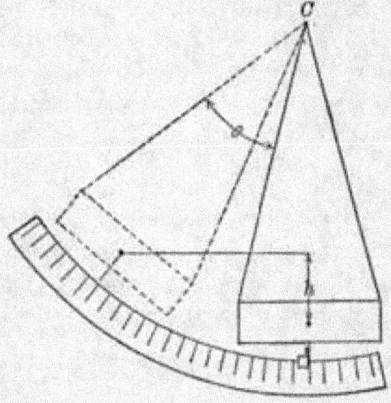

FIG. 34.

The conditions necessary to fulfill the requirements of this equation are met by the use of the ballistic pendulum. This consists (Fig. 34) of a block of wood so suspended that it can swing freely about C as an axis. When a bullet strikes the pendulum bob, the whole impulse may be used in giving to the bob a motion of translation in the direction in which the bullet was moving, or part of the impulse may be used in producing torques which tend to set up wobbling motions that are not taken into account in the above equations. If the bullet strikes at a point called the *center of percussion*, these torques are not produced. The center of percussion is at a distance from the axis of rotation equal to the length of the equivalent simple pendulum, and when the masses of the supporting cords are small compared with that of the bob, the lower end of this equivalent simple pendulum is very near the center of mass of the bob.

If the angle through which the pendulum is deflected by the impact of the bullet is denoted by θ, the height through which the center of mass of the pendulum is elevated by h, and the distance from the axis of rotation to the center of percussion by l, then $h = l(1 - \cos\theta)$. By the time the bullet has ceased to move through the pendulum bob they both have a speed U', and consequently kinetic energy equal to $\frac{1}{2}(m + M)U'^2$. When the end of the swing is reached, this kinetic energy has all been used in lifting them through the distance h; i.e., in doing work equal to $(m + M)gh$.

Consequently, $\quad \frac{1}{2}(m + M)U'^2 = (m + M)gh$.

Whence, $\quad U' = \sqrt{2\,gh} = \sqrt{2\,gl\,(1 - \cos\theta)}$.

On substituting in (68) this value for U', we obtain

$$u = \frac{m + M}{m}\sqrt{2\,gl\,(1 - \cos\theta)}. \tag{69}$$

MANIPULATION. — In setting up the apparatus see that the line of flight of the bullet is horizontal, that it is perpendicular to the axis of rotation of the pendulum, and that it passes through the center of percussion of the pendulum. Weigh the wooden plug in the center of the pendulum bob both before and after the bullet is fired into it. Weigh the rest of the bob, measure l, and observe θ.

Exp. 15. Determination of the Coefficient of Kinetic Friction between a Pulley and an Unlubricated Shaft

THEORY OF THE EXPERIMENT. — Read Art. 22. Consider a shaft of radius r on which rotates a pulley of radius R supporting a string to the ends of which are applied weights W_1 and W_2, Fig. 35. Let

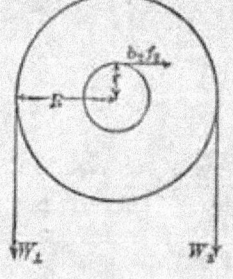

Fig. 35.

the difference between W_1 and W_2 be such that after the pulley is started, it will rotate at a uniform rate. The torque $(W_1 - W_2)R$ is then equal to the opposing torque due to the

friction between the shaft and the pulley. Denoting the coefficient of kinetic friction of the pulley on the shaft by b_2, and
the force pressing the pulley against the shaft by f_2, then the
torque due to friction equals $b_2 f_2 r$. Whence,

$$(W_1 - W_2) R = b_2 f_2 r.$$

If the weight of the pulley be w, then when motion is uniform,
$f_2 = (W_1 + W_2 + w)$, and consequently,

$$b_2 = \frac{(W_1 - W_2) R}{(W_1 + W_2 + w) r}. \tag{70}$$

MANIPULATION. — Clean the shaft and bearing. Weigh the
pulley. By means of calipers and scale, find the diameter of the
pulley and of the shaft. Hang a cord over the pulley and apply
to each end a load of 0.5 kg. Find the overload necessary to
maintain uniform motion when the pulley is once started. From
these data compute the coefficient of kinetic friction for the given
load.

Likewise, find the coefficient of kinetic friction for loads of
about 2 kg., 3 kg., and 4 kg.

Plot a curve having values of $(W_1 + W_2 + w)$ as abscissæ, and
$(W_1 - W_2)$ as ordinates.

Choosing some convenient value of $(W_1 + W_2 + w)$, find from
the curve the corresponding value of $(W_1 - W_2)$, and calculate
the horse power that would be absorbed by friction if the pulley
were making 200 r.p.m.

Exp. 16. Determination of the Coefficient of Kinetic Friction between a Lubricated Journal and its Bearings

THEORY OF THE EXPERIMENT. — The object of this experiment is to determine the coefficient of kinetic friction between a
cylindrical journal and its bearings for different loads, speeds,
and temperatures. The apparatus consists of a spindle passing
through a bearing B (Fig. 36), forming part of a yoke C. The
spindle can be rotated at various speeds by means of a motor,
and the yoke can be loaded by means of adjustable masses M' and
M''. As the spindle is rotated the friction between it and the
bearing tends to rotate the yoke also. This tendency to turn is

measured by the spring dynamometer D, which is essentially a spring balance. By means of a current-carrying conductor in the collar A, the temperature of the oil-covered surface can be con-

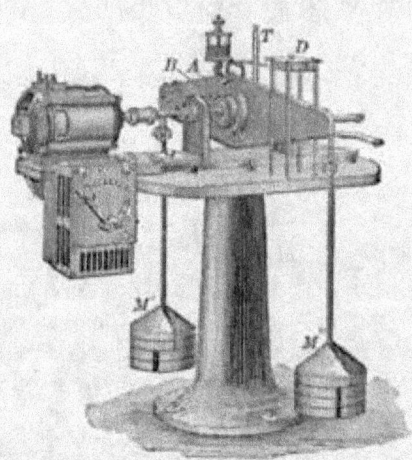

FIG. 36.

trolled. The temperature at any time is indicated by a thermometer T.

If r be the radius of the shaft and F' the total force of friction

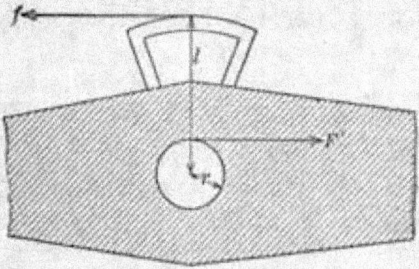

FIG. 37.

tangential to the surface of the shaft, the turning moment resulting from the friction of the shaft and bearing is $F'r$. If f represents the force, having a lever arm l, required to keep the yoke from turning (Fig. 37), the resisting torque is fl. If the center of mass of the yoke with its appendages is vertically below the axis of rotation of the shaft, then when the shaft is rotating and the yoke is held steady,

$$F'r = fl.$$

If the total weight on the bearing surface due to the yoke and its accessories together with the masses M' and M'' be denoted by F, and the coefficient of kinetic friction between the shaft and bearing by b, then, Art. 22, $F'' = bF$. Whence,

$$b\left[=\frac{F'}{F}\right] = \frac{fl}{Fr}. \tag{71}$$

MANIPULATION. — Measure $(l + r)$ and $2\,r$ with calipers and scale. After cleaning the journals and bearing with gasoline, lubricate with the assigned oil and apply small and nearly equal loads to the ends of the arms of the yoke. The difference between these two loads should be sufficient to develop a turning moment due to gravity slightly greater than that due to friction. Start the motor, and by means of the spring dynamometer D measure the tendency of the yoke to turn. Reverse the direction of rotation and take another dynamometer reading. By this operation the pull developed by the friction between the shaft and bearing is first added to the pull on the dynamometer due to the excess weight on one end of the yoke, and then subtracted from it. The difference between the two dynamometer readings is $2\,f$. The data are now at hand for computing the coefficient of kinetic friction between the given surfaces lubricated by the assigned oil, for the particular speed, temperature, and load used in this determination.

The speed can be obtained by means of a stop watch and revolution counter. The values of speed and temperature are not required in the computation, but are required in specifying the conditions under which the determined value of the coefficient of friction applies. Proceeding as above described and keeping the temperature constant, find the coefficient of kinetic friction for various values of F for each speed given by the three steps of the cone pulley.

Plot curves coördinating coefficient of friction and load for each speed, and also curves coördinating coefficient of friction and speed for each load.

Care should be exercised that the direction of rotation of the journal is frequently reversed, especially when the bearing is heavily loaded, so as to avoid error due to inequality of the wearing of the bearing.

Exp. 17. Determination of the Coefficient of Kinetic Friction between Two Plane Surfaces

THEORY OF THE EXPERIMENT. — The object of this experiment is to determine the coefficient of kinetic friction for iron upon iron. From the definition (Art. 22),

$$b = \frac{F_p{'}}{F_n},\qquad (72)$$

where $F_p{'}$ is the force parallel to the slipping surface necessary to maintain uniform motion, and F_n is the force normal to the slipping surface pressing the two bodies together.

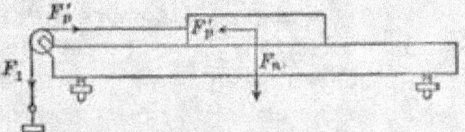

FIG. 38.

The apparatus consists of a horizontal plate having a small pulley fastened at one end, and a block that can be drawn along the length of the plate by means of a cord passing over the pulley, Fig. 38.

Since the slipping surface is horizontal, F_n equals the weight of the block. The value of $F_p{'}$ parallel to the slipping surface required to maintain uniform motion will now be deduced.

Let the weight of the mass on the end of the cord necessary to maintain uniform motion be F_1. On account of the unavoidable friction of the pulley on its bearing, F_1 must be greater than $F_p{'}$. Thus, $F_1 - F_p{'}$ produces a torque about the axis of the pulley, in the direction opposite to the torque produced by the friction of the bearing. When the motion is uniform these torques are equal. Then, if the radius of the pulley be R, that of the shaft r, the resultant force acting on the shaft f_2, and the coefficient of friction between shaft and journal b_2, we may write

$$(F_1 - F_p') R = (f_2 b_2) r$$

or
$$\frac{F_1 - F_p'}{f_2} = \frac{b_2 r}{R} = c \text{ (a constant).}$$

In case the weight of the pulley is sufficiently small compared with the resultant of F_1 and F_p' it may be neglected without affecting the final result more than the unavoidable errors in

FIG. 39.   FIG. 40.

measurement. In this case, the f_2 in the above equation is represented by the distance AB in Fig. 39. Then,

$$c = \frac{F_1 - F_p'}{\sqrt{F_1^2 + F_p'^2}}. \tag{73}$$

Squaring and clearing of fractions,

$$c^2 F_1^2 + c^2 F_p'^2 = F_1^2 - 2 F_1 F_p' + F_p'^2,$$
$$F_p'^2 (c^2 - 1) + F_p' (2 F_1) + F_1^2 (c^2 - 1) = 0,$$

$$F_p' = \frac{-2F_1 \pm \sqrt{4 F_1^2 - 4(c^2 - 1)^2 F_1^2}}{2 (c^2 - 1)} = \frac{F_1 [1 \pm \sqrt{1 - (c^4 - 2c^2 + 1)}]}{1 - c^2}$$

$$= \frac{F_1 [1 \pm \sqrt{c^2 (2 - c^2)}]}{1 - c^2} = \frac{F_1 [1 \pm c \sqrt{2 - c^2}]}{1 - c^2}. \tag{74}$$

It will be recalled that the constant c equals the ratio of the tangential force required to keep the pulley in uniform motion, to the force pressing the pulley against the bearing. This ratio can be most readily obtained from a supplementary experiment in which a cord with a body on each end is hung over the pulley, Fig. 40. The weights of the two bodies must be so adjusted that after the pulley is started, it will rotate with uniform speed. If these weights be W_1 and W_2, the tangential force just sufficient to overcome the friction of the pulley is $W_1 - W_2$, and, the weight of the pulley being neglected, the force pressing the pulley against its bearing is $W_1 + W_2$. Then,

$$c = \frac{W_1 - W_2}{W_1 + W_2} = \frac{A}{B},$$ (75)

where A is used as an abbreviation for $W_1 - W_2$, and B for $W_1 + W_2$.

Making this substitution in (74), we obtain

$$F_p' = \frac{F_1\left[1 \pm \dfrac{A}{B}\sqrt{\dfrac{2B^2 - A^2}{B^2}}\right]}{\dfrac{B^2 - A^2}{B^2}} = F_1\left[\frac{B^2 \pm A\sqrt{2B^2 - A^2}}{B^2 - A^2}\right].$$ (76)

Since the quantity in square brackets is the same for all values of F_p' and F_1, it can be denoted by the single letter k and the above equation written in the abbreviated form

$$F_p' = kF_1.$$

The coefficient of kinetic friction between the two surfaces is, then,

$$b\left[= \frac{F_p'}{F_n}\right] = \frac{kF_1}{F_n}.$$ (77)

MANIPULATION. — After cleaning the block and the surface of the plate and making the plate horizontal with the aid of a spirit level, place the block near one end and add masses to the pan until the block on being started keeps in uniform motion. Make not less than three determinations of F_1 for the given load F_n. Even with carefully machined and hand scraped surfaces the various values of F_1 will differ considerably. In the same manner, find the average value of F_1 when the block is loaded with 100 gm., 300 gm., 500 gm., and 700 gm.

Hang over the pulley a string with the block suspended on one end and an equal mass on the other. Add to one end such an overload that after being started the pulley will move at a uniform rate. From the weights on the two ends of the string find A and B. With these values of A and B compute the constant k.

For each load F_n calculate the value of F_p' and also the coefficient of kinetic friction b.

Plot a curve showing the relation between F_n and F_p'. This curve should be very nearly a straight line, and, if the normal forces, F_n, are plotted as abscissæ, (77) shows that the slope of the curve gives the coefficient of friction. Determine the slope of the curve and see how the result checks with the mean of the previous results.

Exp. 18. Determination of the Moment of Inertia of a Rigid Body by the Rotation Method

THEORY OF THE EXPERIMENT. — Read Art. 23. The comparison of moments of inertia of small bodies is readily effected by means of the apparatus shown in Fig. 41. This consists of a light

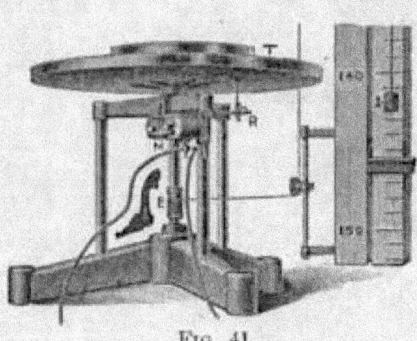

circular table T attached to the upper end of a vertical spindle which rotates with very slight friction. The lower end of the spindle is provided with a drum E, to which a constant torque can be applied by means of a weight acting through a flexible cord arranged as shown. By means of an

FIG. 41.

electromagnetic release MR in electric connection with a switch and a seconds pendulum, the circular table can be released at any clock beat, and thereafter the armature M will give a sharp click at each succeeding second.

If the mass A falling in front of the vertical scale has a uniform linear acceleration a, then the rotating system has a uniform angular acceleration a. If the falling body starts from rest, then in time t it will fall through some distance x, such that $x = \frac{1}{2} a t^2$.

If the spindle on which the cord is wound has a radius r, and the cord neither slips nor stretches, the angular acceleration of the rotating system is

$$a\left[=-\frac{a}{r}\right]=\frac{2x}{rt^2}. \tag{78}$$

Whence, the angular acceleration of the rotating body is directly proportional to the distance x passed over by the falling mass during any assigned time t.

If the torque due to the weight of the falling mass acting on the spindle of radius r be denoted by L, and the moment of inertia of the rotating system with respect to the axis of rotation be denoted by K, then from (21) and (78), the angular acceleration equals

$$\frac{L}{K}=\frac{2x}{rt^2}.$$

Whence,
$$x=\left(\frac{Lrt^2}{2}\right)\frac{1}{K}.$$

All the quantities within the parenthesis are constants. Putting c for the term within the parenthesis,

$$x=\frac{c}{K}.$$

Consequently, if the rotating body be acted upon by a constant torque for a given time, the falling body will pass through a distance inversely proportional to the moment of inertia of the rotating body.

Representing the moment of inertia of the empty table and spindle by K_e, and the distance the falling body passes through in the given time by x_e, we have

$$x_e=\frac{c}{K_e}. \tag{79}$$

Similarly, if there be placed on the table a body whose moment of inertia with respect to the axis of rotation is K_1, and the distance traversed by the falling body in the given time be represented by x_1, we will have

$$x_1 = \frac{c}{K_e + K_1}. \tag{80}$$

And if another body be substituted for this body,

$$x_2 = \frac{c}{K_e + K_2}. \tag{81}$$

By eliminating from (79), (80), and (81) the two unknown quantities c and K_e, we obtain for the ratio of the moments of inertia of the two bodies placed on the rotating table,

$$\frac{K_1}{K_2} = \frac{x_2(x_e - x_1)}{x_1(x_e - x_2)}. \tag{82}$$

Thus, by noting the distance traveled by the falling body in a given time when the rotating table is empty, and then when loaded by any two bodies, one after the other, we obtain the ratio of the moments of inertia of the two bodies with respect to the axis of rotation. By using one body of such simple geometrical shape that its moment of inertia can be computed, the moment of inertia of any other body can be determined.

MANIPULATION. — The above discussion assumes that there is no effect due to friction of the bearings of the rotating shaft or of the pulleys that carry the string. The effect of friction will be neutralized when such a weight is applied to the end of the string that the spindle will rotate with constant angular velocity. First, see that the armature of the magnetic release in the clock circuit beats properly when the switch is closed. Place on the circular table a body of known moment of inertia, for example, a circular ring of known mass and dimensions. Adjust the mass in the scale pan on the end of the string till it falls through equal distances in equal times. Then add 100 gm. to the scale pan and note the distance that it falls from rest in 50 sec.

Substitute for the body of known moment of inertia the body whose moment of inertia is to be determined, and proceed as before.

Weigh and measure the circular ring, and compute its moment of inertia by means of (27).

These data substituted in (82) give the required moment of inertia of the body under investigation, with respect to the axis of rotation.

Exp. 19. Determination of the Moment of Inertia of a Rigid Body by the Vibration Method

THEORY OF THE EXPERIMENT. — Read Art. 23. In this experiment the moment of inertia of a body is to be determined by experiment and also by computation.   In any case where a body can be set into free simple harmonic motion of rotation about the axis with reference to which the moment of inertia is required, it is a simple matter to determine experimentally the moment of inertia of the body.   If the body be of simple geometric shape, the moment of inertia can also be computed.

In elementary dynamics it is shown that a body of moment of inertia K when acted upon by a torque L proportional to the angular displacement from the equilibrium position, vibrates with simple harmonic motion of rotation with the period

$$T = 2\pi\sqrt{-\frac{K\phi}{L}},$$

where ϕ is the amplitude of vibration.

Whence,
$$K = \left(-\frac{L}{\phi(2\pi)^2}\right)T^2.$$

Since the ratio of torque to displacement is constant, the quantity within the parenthesis is constant.   Representing it by the symbol c, the above equation may be written

$$K = cT^2. \tag{83}$$

That is, in the case of any rigid body or system of bodies vibrating freely with simple harmonic motion of rotation, the square of the period of vibration is directly proportional to the moment of inertia of the body.

MANIPULATION. — A convenient form of apparatus for this experiment consists, Fig. 42, of two horizontal disks connected

by three thin vertical rods. From the center of the upper disk rises a short spindle for attachment to the supporting torsion wire. The body whose moment of inertia is required can be placed on the lower disk in such a position that the line about which its moment of inertia is to be determined coincides with the axis of the supporting wire. The positions of the masses MM are then adjusted until the axis of vibration of the system passes through the center of the two disks. Below the vibrating system is a device by means of which the apparatus can be set into torsional vibration with very little swinging motion.

FIG. 42.

If the moment of inertia of the apparatus with respect to the axis of vibration be K_1 and the period be T_1, then (83),

$$K_1 = cT_1^2. \tag{84}$$

Now add to the apparatus a body which has a moment of inertia K_2 with respect to the same axis. If the period of vibration now be T_{12}, the moment of inertia of the system is

$$K_1 + K_2 = cT_{12}^2. \tag{85}$$

Combining (84) and (85) by eliminating c, we find the moment of inertia of the apparatus to have the value

$$K_1 = K_2 \left(\frac{T_1^2}{T_{12}^2 - T_1^2} \right). \tag{86}$$

Now substitute for the body of known moment of inertia the body whose moment of inertia is required, and find the period of vibration as before. If this period be denoted by T_{13}, we find by the method of the preceding paragraph that the moment of inertia K_3 of the body under investigation is

$$K_3 = K_1\left(\frac{T_{13}^2 - T_1^2}{T_1^2}\right).$$

Whence, substituting the value of K_1 from (86), the moment of inertia of the body under investigation is found to have the value

$$K_3 = K_2\left(\frac{T_{13}^2 - T_1^2}{T_{12}^2 - T_1^2}\right). \tag{87}$$

In finding each of these required periods of vibration, first set the pointer P directly in front of one of the three vertical rods while the apparatus is at rest. Then set the apparatus into torsional vibration with an amplitude of about 90°. Be careful that there is no translational motion. At same instant when the given vertical rod passes the pointer start a stop watch. Count 50 complete vibrations and stop the watch. From these data compute the time of one vibration.

Take all of the required linear dimensions with a vernier caliper and make all weighings with a balance of moderate sensitivity.

Compute the moment of inertia, (a) from the observed periods of vibration, and (b) from the mass and dimensions of the given body.

Exp. 20. Determination of the Tensile Coefficient of Elasticity, or Young's Modulus, by Stretching

THEORY OF THE EXPERIMENT. — Read Art. 24. From the definition of Young's modulus, it follows that if L denotes the length of a wire, d its diameter, and e the elongation produced by a force F, then the Young's modulus of the material composing the wire is, (28),

$$E = \frac{4F}{\pi d^2} \div \frac{e}{L} = \frac{4FL}{\pi d^2 e}. \tag{88}$$

If the force is measured in dynes and the other quantities in centimeters, the value of E will be in dynes per sq. cm. The object of this experiment is to determine the value of Young's modulus for a metal in the form of a wire.

Of the quantities which have to be measured, the only one that

it is difficult to get with moderate accuracy is the value of the elongation e. One means of finding this is by an optical lever. The upper end of the wire is securely clamped to a rigid support (Fig. 43), and to the lower end of the wire is fastened a rectangular piece of metal S terminating in a hook for the attachment of a weight pan H. This rectangular piece of metal is kept from twisting or swinging by being let through a loosely fitting rectangular hole in a second bracket fastened to the wall. One leg of the optical lever is supported in the axis of the wire by the rectangular hook, while the other two legs are supported by the bracket.

In Fig. 44, mnb is the optical lever with its mirror vertical, o is a horizontal telescope, and oo' is a vertical scale divided into centimeters and millimeters. If the wire be stretched by a small amount, the optical lever will assume the position $m'nb'$ making an angle θ with its previous position. When light is reflected from a mirror, the angle of reflection equals the angle of incidence. Whence $o'a'i = oa'i = \theta$. Consequently $oa'o' = 2\theta$. The elongation e is the vertical distance through which the point m moves in passing to the position m'. So that

FIG. 43.

$$e = m'n \sin \theta = mn \sin \theta.$$

Now since the small distance aa' is negligible in comparison with ao,

$$\tan 2\theta \doteq \frac{oo'}{ao}.$$

When θ is small and is expressed in radians

$$\tan 2\theta \doteqdot 2\theta$$

and

$$\sin \theta \doteqdot \theta.$$

That is,

$$e \left[= mn \sin \theta \right] \doteqdot mn \, \theta \doteqdot \frac{mn \cdot oo'}{2 \, ao}.$$

On putting this value of e in (88) it becomes

$$E \doteqdot \frac{8 \, FL}{\pi d^2 \cdot mn} \cdot \frac{ao}{oo'}. \qquad (89)$$

MANIPULATION. — See that the wire is straight and carefully suspended. Place three or four kilograms on the supporting

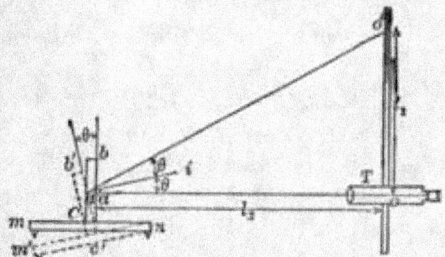

FIG. 44.

bracket directly over the clamp holding the upper end of the wire, and one kilogram on the pan below. Put the optical lever in place and the telescope and scale a meter or so from it, clamp the scale vertical, and adjust the height of the telescope until it is at about the same level as the optical lever. Move the head to such a position that the image of the telescope is seen in the middle of the mirror of the optical lever. If the eyes are not now at the level of the telescope, turn the thumb screw beneath the front legs of the optical lever until the image is seen when the eyes are at the same level as the telescope. This makes the mirror vertical. Focalize the telescope as directed in Art. 14.

Read the telescope, move all the masses from the supporting bracket down to the weight pan, read the telescope, move the masses back to the supporting bracket, and read the telescope again. If the elastic limit has not been exceeded, the last reading should be about the same as the first. Repeat two or three times. Make about five determinations, each one after moving the telescope and scale a few centimeters farther from the optical lever.

Measure the diameter of the wire in some half dozen places with a micrometer caliper. Determine the length mn of the optical lever by pressing the three feet upon a piece of cardboard, connecting the prick points made by the two front feet by a fine line, and then measuring the normal distance between the remaining prick point and this line by means of a millimeter scale. Determine the length of the wire with a meter stick, and the loads added to the weight pan with a platform balance weighing to grams.

For each distance ao find the average deflection oo' and calculate $\dfrac{ao}{oo'}$. Find the average of all the values for $\dfrac{ao}{oo'}$, and by (89) calculate E. Give the result in dynes per sq. cm., in Kg. wt. per sq. mm., and in lb. wt. per sq. in.

Exp. 21. Determination of the Tensile Coefficient of Elasticity or Young's Modulus by Bending

THEORY OF THE EXPERIMENT. — Read Art. 24. Consider a rectangular rod of length L', breadth B, and depth D, fixed at one end and weighted at the other. The rod will become bent as in the figure. The upper portion of the rod is extended and the lower portion compressed. Since the rod is strained by a longitudinal stress, and since Young's modulus is defined as the ratio of the longitudinal stress to the longitudinal strain, Young's modulus may be determined from an observation of the amount of bending which a given force produces in the rod. The object of this experiment is, by the method of bending, to determine the Young's modulus of the material composing a rectangular rod.

Imagine the unstrained rod to be divided into laminæ by a series of planes normal to its length. Then let the rod be slightly bent by a force F' applied downward at the end of the rod, and let some lamina *abcd* be thereby so distorted that its sides *ad* and *bc* make with each other a small angle $d\theta$. The restoring stress in this lamina produces a couple which tends to bring the rod back to its undistorted position, and is prevented from doing so only by the moment of the distorting force F'.

The first step in the development of the formula for determining the Young's modulus of the rod is to find an expression for the restoring couple due to the stress in this lamina. Halfway between the upper and the lower surfaces of the rod is a neutral surface *gh* which is neither extended nor compressed. Any layer above this surface and parallel to it will be extended, while one below will be compressed. Consider the thin layer *vy* at a distance z from the neutral surface, which has a depth dz and breadth B equal to that of the rod. The original length of this layer was dx. From the figure its elongation is seen to be $z\,d\theta$.

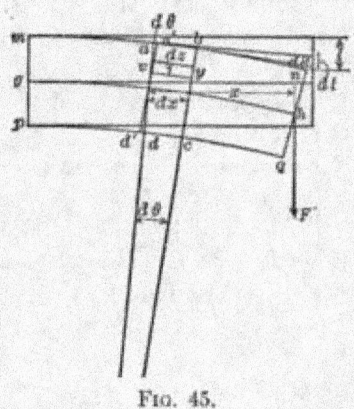

FIG. 45.

The cross section is $B\,dz$. From (28) the force of restitution in the direction of the length of the rod developed in the layer *vy* is

$$\left[\frac{\text{Young's modulus} \times \text{cross section} \times \text{elongation}}{\text{original length}}\right] = \frac{E \times B\,dz \times z\,d\theta}{dz}.$$

Since the lever arm of this force is z, the torque, *i.e.*, the moment of the force of restitution developed in this layer, is

$$\frac{EB \cdot d\theta \cdot z^2\,dz}{dx}.$$

The restoring torque developed by the straining of all the layers above the neutral surface is the sum of the torques developed in the various layers. Thus, integrating the above expression between the limits $z = 0$ and $z = \frac{1}{2} D$, the restoring torque due to all the layers above the neutral surface is seen to be

$$L = \frac{EB \cdot d\theta \cdot D^3}{24\,dx}.$$

Now the resultant moment of the restoring forces below the neutral surface equals the moment of those above. It follows that the whole torque due to the strain in the given lamina is $2\,L$. Since the bar is in equilibrium, this restoring couple equals the distorting moment of F' about e. If the rod be bent only slightly, the moment of F' about e is so little smaller than $F'x$ (Fig. 45) that we may write

$$\frac{EBD^3\,d\theta}{12\,dx} = F'x. \tag{90}$$

The next step is to find the depression l of the end of the rod in terms of $d\theta$. At a and b draw two lines tangent to the curved surface of the rod and equal in length respectively to the arcs an and bn. Then the angle between these tangents equals the angle $d\theta$ between ad and bc. Denoting the depression of the end of the rod due to the bending of the given lamina by dl, we can write

$$d\theta = \frac{dl}{x}.$$

Substituting this value in (90) we have for the depression of the end of the rod due to the distortion of the given lamina $abcd$,

$$dl = \frac{12\,F'}{EBD^3}\,x^2\,dx. \tag{91}$$

The depression l of the end of the rod due to the distortion of all the laminæ is the sum of the depressions due to the separate laminæ. Thus, integrating (91) between the limits $x = 0$ and $x = L'$, we have for the total depression of the end of the rod

$$l = \frac{4\,F'L'^3}{EBD^3}. \tag{92}$$

If the rod, instead of being fastened at one end and loaded at the other, is supported on two knife edges and loaded in the middle, the bending is practically the same as if it were fastened at its middle point and had acting upward upon it at each end a force half as great as the load actually applied. Let the distance between the knife edges be $L = 2 L'$, and the force applied be $F = 2 F'$. Then on substituting for L' and F' in (92), we get

$$l = \frac{FL^3}{4 EBD^3}$$

or

$$E = \frac{F}{4 BD^3} \cdot \frac{L^3}{l}. \tag{93}$$

It can be shown that the above approximate result is accurate to within about 0.03 per cent if the depression of the rod is not more than one hundredth as great as the length of the rod. Accuracy so great as this is seldom required in a determination of Young's modulus, and, besides, the error introduced by the method of measurement is usually greater than that due to the approximation.

MANIPULATION. — Measure B and D at a number of points along the rod by means of a micrometer caliper. Measure L, the distance between the two knife edges, with a meter stick. Place the rod on the knife edges and suspend from the middle point a pan containing sufficient load to bring the rod into good contact with the knife edges. The flexure l of the rod produced by an additional load F may be measured by means of a microscope fitted with an eyepiece micrometer, or by means of a micrometer screw placed above the center of the rod and moving in a nut fastened to a rigid support.

If an eyepiece micrometer be employed, Fig. 46, it must first be standardized as described in Art. 12. The microscope is then focalized on the end of a needle attached to the saddle that supports the binding load. The length of the microscope tube must not be altered after the eyepiece micrometer has been standardized.

If an ordinary micrometer screw be used, Fig. 47, the instant when the screw comes into contact with the rod can be determined either by means of a telephone in a battery circuit including the rod and micrometer screw, or by observing the image of some

FIG. 46.

fixed object in a small mirror one end of which rests upon the rod while the other end rests upon an adjacent fixed support.

Some micrometer screws are provided with a ratchet which causes the head to slip when the end of the screw presses against any object. With such a micrometer screw no device is necessary to indicate when the screw is in contact with the rod.

FIG. 47.

By means of either an eyepiece micrometer or a micrometer screw, take a reading when the rod is not loaded and again when loaded. Remove the load and read again. The amount of load to be used will depend upon the size of the rod. Take similar readings for three different loads. Repeat twice, both to be sure that the elastic limit has not been exceeded and to get a number of determinations of the flexure. Then alter by a few centimeters the distance between the knife edges, and repeat. Take three different lengths, and for each length, using the average flexure for that length, calculate the ratio $\frac{L^3}{l}$. Find the average of the five values of $\frac{L^3}{l}$, and by (93) calculate the Young's modulus of the rod. Express the result in dynes per sq. cm., Kg. wt. per sq. mm., and lb. wt. per sq. in.

Exp. 22. Koenig's Method for the Determination of Young's Modulus of a Rod by Bending

THEORY OF THE EXPERIMENT. — Read Exp. 21. If the rod be so stiff that the depression of the middle point is small, the depression can be determined by the following method with a greater degree of precision than by the direct methods used in the previous experiment.

Consider a straight rectangular rod of breadth B and depth D resting upon two knife-edges distant from one another by a length L. When a force F is applied perpendicularly to the length of the rod at a point midway between the two supports, the rod is bent into a curve which is approximately a circle

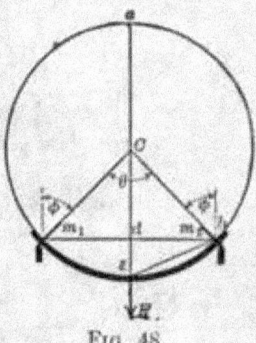

Fig. 48.

so long as the bending is not excessive. In Fig. 48, the angle θ between the normals to the rod at the points of support has the value

$$\theta = \frac{L}{r}, \tag{94}$$

where r is the radius of the circular arc into which the rod is bent. In the triangles abe and dbe

$$\frac{(ae)}{(eb)} = \frac{(eb)}{(de)}$$

or, denoting the deflection de of the middle point of the rod by l,

$$\frac{2\,r}{(eb)} = \frac{(eb)}{l},$$

$$2\,rl = (eb)^2 = \left(\frac{L}{2}\right)^2,$$

$$r = \frac{L^2}{8\,l}.$$

On substituting this value in (94)

$$\theta\left[=\frac{L}{r}\right]=\frac{8l}{L}$$

or

$$l=\frac{\theta L}{8}.$$

On substituting this value in (93) we obtain

$$E\left[=\frac{FL^3}{4\,BD^3l}\right]=\frac{2\,FL^2}{\theta BD^3}. \tag{95}$$

The value of θ can be easily determined by experiment. If two mirrors be fastened rigidly to the rod directly above the supports, then when a force is applied to the middle point, the inclination of the mirrors to one another will change by an amount equal to the change of the angle θ between the normals to the rod at the points of support.

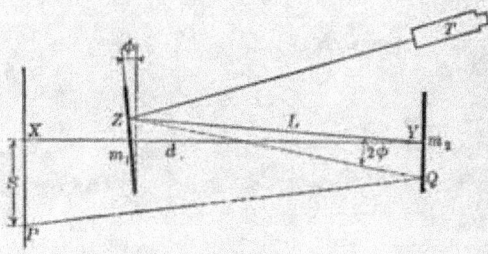

FIG. 49.

Let a telescope and a vertical scale be placed facing the mirrors as shown in Fig. 49. Let the deflection of the mirror m_1 produced by the bending of the rod be denoted by ϕ, the distance between the two mirrors by L, and the distance from the scale to the mirror m_2 by d.

When the mirrors are in the position of the figure, the light that enters the telescope traverses the path $XYZT$. If the mirror m_1 be tilted through the angle ϕ while the mirror m_2 remains stationary, the light that enters the telescope travels the path

PQZT. Since when a mirror is tilted through a certain angle, the reflected ray is tilted through twice that angle, the angle *QZY* $= 2\phi$.

From the figure,

$$\tan 2\phi = \frac{s}{L+d},$$

where s is the difference of scale reading due to the tilting of m_1. Since the angle 2ϕ is always very small, we may write

$$s \doteq 2\phi(L+d).$$

But since a deflection of the rod causes a tilting of both mirrors through the same angle ϕ, the total change in the scale reading, S, is

$$S \doteq 2\phi(L+d) + 2\phi d \doteq 2\phi(L+2d).$$

From Fig. 48, $\theta = 2\phi$. Consequently,

$$\theta \doteq \frac{S}{L+2d}.$$

On substituting this value in (95), we obtain

$$E \doteq \frac{2L^2(L+2d)}{BD^3}\left(\frac{F}{S}\right). \tag{96}$$

MANIPULATION. — Measure B and D at a number of points along the rod by means of a micrometer caliper. Measure L, d, and S with a meter stick. Find the flexure S for three different loads F. Calculate Young's modulus for each load. Express the result in dynes per sq. cm., Kg. wt. per sq. mm., and lb. wt. per sq. in.

Exp. 23. Determination of the Simple Rigidity of a Wire, or Rod, by the Static Method

THEORY OF THE EXPERIMENT. — Read Art. 24. The object of this experiment is to determine the simple rigidity of a cylindrical wire or rod.

Consider a cylindrical rod or wire of length l and radius r with one end fixed and the other end twisted through an angle ϕ. This will cause an element of the surface as AB to be displaced to AB'. From the diagram the shearing strain in the outside layer of the cylinder is $\dfrac{BB'}{l}$. And since BB' $= \phi r$, it will be seen that at every point of the wire distant r_1 from the axis and l from the fixed end, there is a shearing strain equal to $\dfrac{\phi r_1}{l}$. If S denote the shearing stress developed at a point distant r_1 from the axis and l from the fixed end, and μ the simple rigidity of the wire, it follows from the definition of simple rigidity that

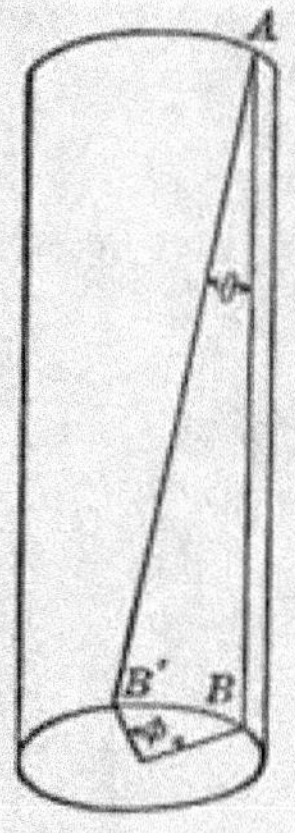

$$\mu = \frac{S}{\dfrac{\phi r_1}{l}}.$$

Fig. 50.

Whence,
$$S = \frac{\mu \phi r_1}{l}. \tag{97}$$

This is the value of the shearing stress at any distance r_1 from the axis of the wire.

The next step is to find what torque would be needed to keep the wire twisted as it is in the figure. Imagine the end of the rod to be divided into n thin concentric rings of width dr_1. Fix the attention for a moment on the ring of radius r_1, Fig. 51. Now since the area of cross section of this elementary ring is $2 \pi r_1 \, dr_1$, the restoring force set up in this particular element has the magnitude

$$\frac{2 \pi \mu \phi r_1^2 \, dr_1}{l}.$$

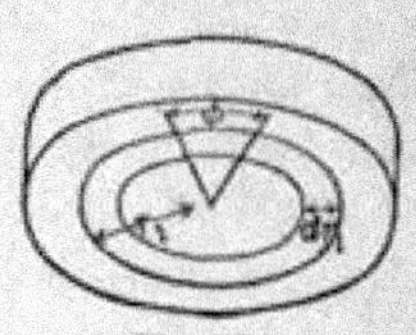

Fig. 51.

Since this force is tangent to a circle of radius r_1, the torque that must act upon this elementary ring to maintain the twist is

$$dL = -\frac{2 \pi \mu \phi r_1^3 \, dr_1}{l},$$

where the negative sign is used because the restoring torque and the displacement are in opposite directions. The total torque developed in all the rings of which the rod is conceived to consist is obtained by integrating the above expression between the limits $r_1 = 0$ and $r_1 = r$. Thus, the torque L that must be applied to the end of the wire to keep it twisted is

$$L = -\frac{\pi\mu\phi r^4}{2l} = -\frac{\pi\mu\phi d^4}{32\,l}, \qquad (98)$$

where ϕ is the angle of twist expressed in radians and d is the diameter of the rod.

In the method here to be employed for determining L, there is fastened to the lower end of the rod a massive disk which has its upper face graduated in degrees, and has around its edge a series of pins placed $20°$ apart. In front of the disk and in back of it are two horizontal scales. The twisting couple is applied to the disk by horizontal forces acting tangentially at its circumference. Equal masses m are suspended by cords which pass in front of the two horizontal scales. Tied to each supporting cord at about the level of the pins in the disk is another short cord which has at its other end a loop that can be slipped over one of the pins, thus twisting the graduated disk through an angle which can be read by means of a pair of pointers fixed above it.

Fig. 52.

Let the forces in the horizontal cords be denoted by F_1 and F_2. Then from the diagram (Fig. 53)

$$\tan w = \frac{h}{x}. \qquad (99)$$

And since F_1 and mg are perpendicular to each other, and F_1, mg, and the tension in the supporting cord are a system of concurrent forces in equilibrium,

$$\frac{mg}{F_1} = \tan w. \tag{100}$$

From (99) and (100) it follows that

$$F_1 = \frac{mgx}{h}.$$

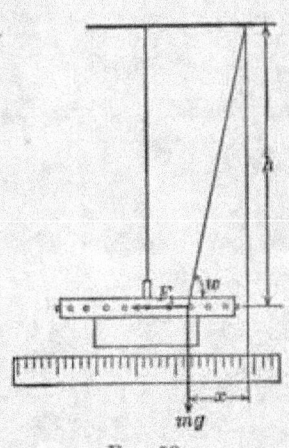

If the horizontal threads are looped over diametrically opposite pins, and if the points from which the upright cords hang are equidistant from the plane of the wire and supporting bracket, $F_1 = F_2$. If we drop the subscripts and denote by D the diameter of the disk increased by twice the radius of the horizontal cords, the moment of the couple that tends to turn the disk farther from its equilibrium position is of the magnitude

FIG. 53.

$$FD = -\frac{mgxD}{h}. \tag{101}$$

When the disk is in equilibrium, the torque L that must be applied to the end of the wire to keep it twisted is equal in magnitude to FD. Whence, from (98) and (101),

$$\frac{\pi\mu\phi d^4}{32\,l} = \frac{mgxD}{h}. \tag{102}$$

Finally, writing in place of ϕ radians its value $\frac{\beta}{360} \cdot 2\pi$ radians,

where β is the number of degrees in ϕ radians, (102) gives

$$\mu = \frac{5760\,g\,Dl}{\pi^2\,d^4h} \cdot \frac{mx}{\beta}. \tag{103}$$

MANIPULATION. — Carefully measure the diameter of the rod or wire in at least ten places with a micrometer caliper. Take the diameter of the disk with a vernier caliper. Measure h and l with a meter stick or steel tape. Use such loads and loop the cords over such pins as to get a series of some half dozen values for β, each somewhat larger than the one before it, but the largest not much more than 90°. The loads in the two pans must be equal, and the cords should be looped over pins far enough around to give fairly large values for x. In getting each end of the distance x, record the reading on each side of the cord and use the mean as being the position of the middle of the cord. Find the average value of $\frac{mx}{\beta}$, and by (103) find μ. Express the result in dynes per sq. cm., Kg. wt. per sq. mm., and lb. wt. per sq. in.

Exp. 24. Determination of the Absolute Coefficient of Viscosity of a Liquid by Poiseuille's Method

THEORY OF THE EXPERIMENT. — Read Art. 25. Consider a column of liquid flowing through a tube of length l, and with a radius, R, so small that there will be no eddies in the liquid column. Imagine this column to be made up of a large number of concentric hollow cylinders of very small thickness Δr. Suppose that all of these hollow cylinders but one could be made solid, so that there would be a solid rod surrounded by a thin layer of the fluid, and this again surrounded by a solid tube. While the rod was moving, two forces would be acting on it — one due to the viscous resistance in the tube that was still liquid, tending to retard the motion of the rod, and the other due to the difference between the pressures at the two ends of the rod, tending to accelerate it. If the radius of the rod were r, the viscous resistance in the liquid tube surrounding it would, by (29), be

$$F_r \left[= \frac{\eta A \, (s_1 - s_2)}{x} \right] = \frac{\eta \cdot 2\pi r l \cdot \Delta s}{\Delta r}.$$

And if p denote the difference between the pressures at the two ends of the rod, the force to which this difference in pressure would give rise would be

$$F_g = p\pi r^2.$$

If the rod were moving uniformly, the magnitude $F_v = F_p$, that is,

$$\frac{\eta \cdot 2\pi r l \cdot \Delta s}{\Delta r} = p\pi r^2$$

or $$ds = \frac{pr \cdot dr}{2\eta l},$$

whence by integrating,

$$s = \frac{pr^2}{4\eta l} + C \text{ (a constant)}.$$

The constant of integration can be readily obtained. When $r = R$, $s = 0$, and

$$0 = \frac{pR^2}{4\eta l} + C,$$

$$\therefore \quad C = -\frac{pR^2}{4\eta l}.$$

Consequently the speed of flow of the given cylinder is

$$s = \frac{pr^2}{4\eta l} - \frac{pR^2}{4\eta l}. \tag{104}$$

We are now in position to find the value of the volume of liquid V, discharged from the tube in time t. The volume of liquid discharged in time t by the cylindrical element of radius r and thickness Δr, moving with speed s, is

$$\Delta V = \Delta A \cdot s \cdot t$$

$$= 2\pi r \, \Delta r \cdot s \cdot t$$

$$= 2\pi r \, \Delta r \cdot t \left(\frac{pr^2}{4\eta l} - \frac{pR^2}{4\eta l} \right),$$

$$\therefore \quad dV = \frac{\pi p t r^3 \, dr}{2 l\eta} - \frac{\pi p R^2 t r \, dr}{2 l\eta}.$$

Integrating, the value of the volume discharged by the entire tube in time t is found to be

$$V = \frac{\pi p t}{2 l \eta} \int_R^0 r^3 \, dr - \frac{\pi p R^2 t}{2 l \eta} \int_R^0 r \, dr.$$

That is,

$$V = \frac{\pi p R^4 t}{8 l \eta}. \tag{105}$$

If the pressure be due to a column of liquid of height h and density ρ, then $p = \rho g h$. On putting this value in (105), and solving for η, we have

$$\eta = \frac{\pi \rho g h R^4}{8 l} \cdot \frac{t}{V} \tag{106}$$

dynes per sq. cm. per unit velocity gradient.

It should be noticed that in deriving (106) it has been tacitly assumed (a) that the viscous resistance to the flow of the liquid is uniform throughout the entire length of the tube, (b) that the lines of flow of liquid in the tube are parallel to the axis of the tube throughout its length, (c) that no part of the energy supplied to the liquid in the tube appears as energy of motion, (d) that there is no effect at the outlet due to surface tension. The conditions demanded by (a), (b), and (c) can be realized to a sufficient degree of approximation by using a tube that is both long and of narrow bore and having the liquid flow through at a uniform rate. Condition (d) is met by immersing the discharge orifice in a portion of the liquid having a considerable free surface.

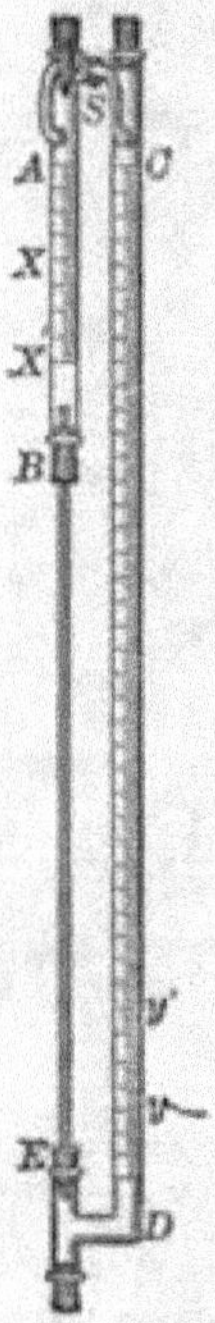

MANIPULATION. — A viscometer that fulfills the above conditions is illustrated in Fig. 54. The vertical tubes AB and CD are of uniform bore and are graduated in millimeters throughout their length. The capillary tube BE is straight and of uniform circular bore. In order that the temperature of the liquid being investigated shall be constant and definite, the viscometer is supported in a suitable water jacket supplied with a thermometer.

Fig. 54.

The length l of the capillary tube is measured with a meter stick. The mean radius of the bore is determined by measuring the length of a known mass of mercury at different positions along the length of the tube. An amount of mercury sufficient to make a thread about four centimeters long is drawn into the tube by suction applied at the opposite end, and this thread is measured in length at different equally spaced positions along the length of the tube by means of a dividing engine. Knowing the mass of the mercury thread and the average length, the average radius of the bore of the tube is determined. A tube with a bore departing very much from uniformity must be rejected in determining the absolute coefficient of viscosity.

In order to determine the V in (106) it is necessary to calibrate the lower part of the tube CD. This may be done by putting a solid stopper at E, removing the one just above C, and dropping into CD known volumes of water from a burette. After each small volume of water is dropped in, a reading is made of the top of each water column — the one in CD and the one in the burette. From these readings a curve is to be plotted coördinating the volume of water in CD with the reading of its surface on the CD scale.

After thoroughly cleaning and drying the parts of the viscometer, it is assembled, a quantity of the liquid under investigation is introduced, and this liquid column run back and forth until it is free of air bubbles and the tubes are coated with a thin film of the liquid. The quantity of liquid introduced should be such that it will form a column extending from a point near the upper end of the tube AB to a point near the lower end of CD.

With all the rubber stoppers tight, and the stopcock S open, run the liquid into the position mentioned above, close the stopcock, and place the viscometer in the water bath. After the temperature has become constant and of the desired value, with stop watch in hand open the cock S, and when the meniscus in AB reaches some previously selected scale division X, start the watch; when the meniscus reaches some second selected scale division X', stop the watch. This gives the value for t in (106).

When the upper meniscus was at X the lower meniscus was at some point y, and when the upper meniscus had fallen to x' the lower meniscus had risen to some point y'. The positions of y and y' can be obtained by opening S and again running the liquid into AB to the points x and x'. The mean of the vertical distances between x and y, and x' and y', is the value for h in (106). These distances can be obtained by the scales engraved on AB and CD. ρ can be obtained by means of a balance and a 5 cc. pipette. V is obtained by finding from the curve already plotted the volume of water that would be held between the marks y and y'.

At least five sets of observations should be taken and the average value for $\dfrac{t}{V}$ used in (106) to get η at the temperature of the experiment.

Exp. 25. Determination of the Absolute Coefficient of Viscosity of a Liquid by the Rotating Liquid Method

THEORY OF THE EXPERIMENT. — Read Art. 25. This method is based upon a determination of the couple required to prevent turning of a cylinder suspended in a rotating coaxial cylinder filled with a sample of the liquid under test.

Let AA' and BB', Fig. 55, represent transverse sections of the two cylinders, the space between them being filled with a sample of the liquid under test. Let the outer cylinder be rotated about the common axis with a constant angular speed w, while the inner cylinder is prevented from rotating by the restoring torque developed in a wire suspending it. Since there is no slip between the cylinders and the liquid in immediate contact with them, and since there is no

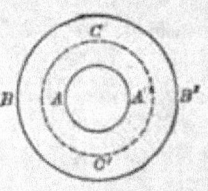

FIG. 55.

abrupt change of velocity between contiguous particles of a liquid, each particle of the liquid must be rotating about the common axis of the cylinders, and the angular speed of the successive layers of liquid must fall off from w at the outer cylinder to zero at the inner cylinder.

Consider the motion of the liquid particles at an imaginary

coaxial cylindrical surface of which CC' represents a right section. The speeds of the particles on the outer side of this surface will be greater than that of the particles on the inner side. Let s' denote the velocity gradient normal to this surface. Then if r be the radius and l the length of this cylindrical surface, its area is equal to $2\pi rl$. And since each unit of area will be acted upon by a tangential stress, (29), equal to $\eta s'$, the torque developed about the axis through O is measured by

$$(\eta s' \cdot 2\pi rl)\, r = 2\pi r^2 l\eta s'.$$

Since the liquid between AA' and CC' is in steady motion, the forces acting on it must be in equilibrium. The cylinder AA' must therefore exert a torque

$$L_e = 2\pi r^2 l\eta s', \tag{107}$$

tending to retard the rotation of the fluid in contact with it; and an equal but oppositely directed torque must be exerted on the stationary cylinder. This reacting torque must be balanced by the restoring torque set up in the suspending wire when the inner cylinder is in equilibrium. If the modulus of torsion τ of the suspending wire be known (i.e., the torque per unit twist), and the twist or angular displacement ϕ, the restoring torque L can be obtained from the relation

$$L = \tau\phi. \tag{108}$$

To obtain a measure of the velocity gradient s', consider two adjacent points on the radius OB equidistant from and on opposite sides of the surface CC'. Let the distances of these points from the axis O be represented by r_a and r_b, and let their angular velocities about the given axis be represented by w_a and w_b, respectively. The difference between the linear speeds of the liquid at these points is $(w_b r_b - w_a r_a)$. This difference may be considered as made up of two parts,

$$(w_b r_b - w_a r_a) = \frac{w_b + w_a}{2}(r_b - r_a) + \frac{w_b - w_a}{2}(r_b + r_a).$$

The first term of the right-hand member expresses a difference in the linear speeds of two points moving with the same angular speed $\frac{1}{2}(w_b + w_a)$. The second term expresses a difference in the linear speed of two points moving with different angular speeds about a common axis. When the angular velocities are the same, there is no angular displacement of one point relative to the other. Consequently no shearing stress is involved in the first term. The difference of linear speed which involves a shearing stress is given by the second term. Hence, from the definition of velocity gradient, (Art. 25),

$$s'\left[=\frac{s_b - s_a}{r_b - r_a}\right] = \frac{w_b - w_a}{r_b - r_a}\frac{(r_b + r_a)}{2} = \left(\frac{w_b - w_a}{r_b - r_a}\right)r,$$

whence, from (107), the torque acting on the cylindrical surface is

$$L_c = 2\pi r^3 l\eta \frac{w_b - w_a}{r_b - r_a}$$

or

$$\frac{2\pi l\eta}{L_c}(w_b - w_a) = \frac{r_b - r_a}{r^3}.$$

Multiplying the numerator of the right-hand member by $\frac{1}{2}(r_b + r_a)$ and the denominator by the equal quantity r,

$$\frac{2\pi l\eta}{L_c}(w_b - w_a) = \frac{1}{2}\frac{r_b^2 - r_a^2}{r^4}.$$

If we now let r_b differ only infinitesimally from r_a, we may write $r^2 = r_a r_b$ and the above equation becomes

$$\frac{4\pi l\eta}{L_c}(w_b - w_a) = \frac{r_b^2 - r_a^2}{r_a^2 r_b^2} = \frac{1}{r_a^2} - \frac{1}{r_b^2}. \tag{109}$$

Now consider the distance AB between the two cylinders, Fig. 55, to be divided into n cylindrical elements of infinitesimal thickness. On setting up an equation of the form of (109) for each of these cylindrical elements extending from $r_0 = OA = R_1$ to $r_n = OB = R_2$, and adding these equations, we obtain

$$\frac{4\pi\eta l}{L_c}(w_n - w_0) = \frac{1}{r_0^2} - \frac{1}{r_n^2}$$

Whence, the torque acting on the cylindrical surface of the suspended cylinder has the value

$$L_c = \frac{4\,\pi\eta l\,(w_n - w_0)}{\dfrac{1}{r_0^2} - \dfrac{1}{r_n^2}}.$$

Substituting for r_0 and r_n their values R_1 and R_2, respectively, and putting $w_n - w_0 = w$, we have

$$L_c = \left(\frac{4\,\pi R_1^2 R_2^2}{R_2^2 - R_1^2}\right) w l \eta.$$

For a particular apparatus, the quantity within the parenthesis is a constant quantity which will be represented by c. Then

$$L_c = cwl\eta. \tag{110}$$

Representing the torque acting upon the ends of the suspended cylinder by L_e, the total torque may be written $L = L_c + L_e$. Whence, (108) and (110),

$$L\,[= \tau\phi] = cwl\eta + L_e. \tag{111}$$

The experiment can be arranged so that the value of L_e need not be determined. The method consists in measuring the couple developed on a cylinder of length l_1 when the container is rotating at a known speed w, and then, with the container rotating at the same speed, measuring the couple developed on a second cylinder exactly like this one except that the length is l_2. Representing the deflections of the two cylinders by ϕ_1 and ϕ_2, respectively, we have for the torques in the two cases,

$$L_1 = \tau\phi_1 = cwl_1\eta + L_e,$$
$$L_2 = \tau\phi_2 = cwl_2\eta + L_e,$$

from which L_e can be eliminated and the coefficient of viscosity obtained.

$$\eta = \frac{\tau\,(\phi_2 - \phi_1)}{(l_2 - l_1)\,cw} \tag{112}$$

dynes per sq. cm. per unit velocity gradient.

MANIPULATION. — In the apparatus used in this experiment, Fig. 56, the containing cylinder can be rotated about its geometric axis at any constant angular speed from 40 to 100 revolutions per minute. The speed can be determined either by a tachometer or a revolution counter and stop watch. The inner cylinder is suspended coaxially with the outer cylinder by means of a thin wire of steel or phosphor bronze. The upper end of the inner cylinder should be two or more centimeters below the surface of the liquid under test.

In the case of a liquid of small viscosity the deflection will be small even though the suspending wire be thin. These small deflections are best measured by means of a mirror m attached to the supporting rod, and a horizontal telescope and scale not shown in the figure. In the case of such viscous substances as glues, clays, and glazes the deflections will be large even when a much larger suspending wire is employed. Under these conditions the mirror, telescope, and scale may well be replaced by a circular scale attached to the supporting rod and a fixed index pointer attached to the frame of the apparatus.

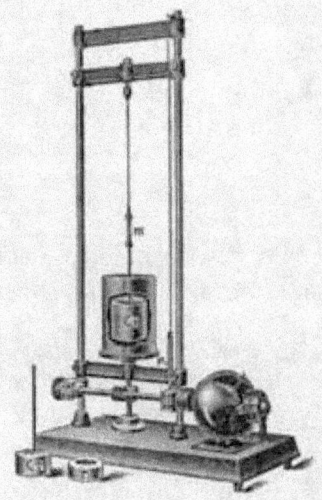

FIG. 56.

By means of a caliper and steel scale find the diameter of the inner cylinder and the inside diameter of the outer cylinder. The value of c is then computed from the expression

$$c = \frac{4 \pi R_1^2 R_2^2}{R_2^2 - R_1^2}. \tag{113}$$

The lengths of the two suspended inner cylinders are measured by means of a steel scale.

The value of the modulus of torsion τ of the suspending wire remains to be found. In elementary dynamics it is shown that the period T of a body oscillating with a simple harmonic motion of rotation is given by the expression

$$T = 2\pi\sqrt{\frac{K\theta}{L}},$$

where K is the moment of inertia of the body with respect to the axis of vibration, and L is the restoring torque developed by an angular displacement θ. In the present case, the restoring torque acting upon the suspended cylinder is due to the rigidity of the suspending wire. Since from (108) $L = \tau\theta$, it follows that

$$T = 2\pi\sqrt{\frac{K}{\tau}} \tag{114}$$

and
$$\tau = \frac{4\pi^2 K}{T^2}. \tag{115}$$

It follows that, if we know the period and the moment of inertia of the suspended cylinder, the value of the modulus of torsion can be computed. Usually, however, the attachment of the wire to the cylinder is by means of a clamp of such shape that the value of K cannot be computed. In this case, the following experimental arrangement will render unnecessary any knowledge as to the value of K.

Set the suspended cylinder into torsional vibration in air and find the period T. Then

$$T = 2\pi\sqrt{\frac{K}{\tau}}.$$

Place on top of the cylinder a ring of known moment of inertia K_1, and find the period of vibration, T_1, of the system now having a moment of inertia $K + K_1$. In this case,

$$T_1 = 2\pi\sqrt{\frac{K + K_1}{\tau}}.$$

Eliminating K from these equations, we have

$$\tau = \frac{4\pi^2 K_1}{T_1^2 - T^2}. \tag{116}$$

After determining the modulus of torsion, τ, of the wire by either (115) or (116), suspend the short cylinder in the liquid under test and measure ϕ_1 for each of several angular speeds w. Replace this cylinder by the long one and record values of ϕ_2 for the same angular speeds. The speed is regulated by adjusting the series resistance in the motor circuit. All of the data are now at hand for substitution in (112).

Care should be taken that the axis of the suspended cylinder and the axis of the containing vessel coincide with the axis of rotation. The bottom of the suspended cylinder should be about two centimeters above the bottom of the containing vessel. In each case the cylinder should be covered with the specimen under test to a depth of about two centimeters.

CHAPTER III

OPTICS

26. Light Units. — The total visible energy emitted per second by a luminous source is called the total *flux of light*. For purposes of comparison, the light source used as a standard is a lamp burning amyl acetate devised by Hefner. The unit of light-flux is the light-flux emitted in one space radian by a Hefner lamp and is called the *hefner lumen*. Since there are 4π space radians in a complete sphere, the total light-flux from the Hefner lamp is 4π hefner lumens.

The *intensity of a point light source* is measured by the light flux emitted per unit solid angle. Thus, if the total flux emitted be F,

$$I = \frac{F}{4\pi}. \qquad (117)$$

This equation shows that the intensity of a point light source will be unity when it emits a flux $F = 4\pi$ hefner lumens. This is the light-flux of a standard Hefner lamp. Consequently, the unit intensity of a point light source is called the *mean spherical hefner*.

The light-flux per unit area of cross section of the beam is called the *light-flux density*, or the *illumination*, of the stream of light. Thus

$$E = \frac{F}{a}. \qquad (118)$$

From this equation we derive a unit of illumination — one hefner lumen of light-flux per square meter of area. This unit is called the *hefner-lux*.

If we have a point source, then at a distance r the illumination would be

$$E \left(= \frac{F}{a} \right) = \frac{F}{4\pi r^2}.$$

Whence, from (117),

$$E = \frac{I}{r^2}.$$ (119)

From this equation we see that at a distance of one meter from a point source of luminous intensity one spherical hefner, the illumination is unity. For this reason, the unit of illumination is also called the *hefner-meter*.

Before the Hefner standard lamp was devised, candles were commonly used as light standards. The standard British candle was one that burned 120 grains of spermaceti per hour. On account of the lack of uniformity of even the most carefully made

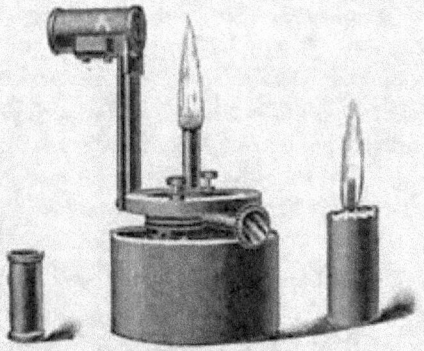

Fig. 57.

candles, they are now seldom used in actual measurements of light quantities. But as actual candles were employed for a long time, light quantities are still often expressed in terms of candles. Thus, luminous flux is expressed in *candle lumens*, luminous intensities in *mean spherical candles*, and illuminations in *candle-lux* or *candle-feet*.

In transforming light quantities expressed in terms of the Hefner lamp into the corresponding quantities expressed in terms of the standard candle, it has been agreed to take ten-ninths spherical hefners to equal one spherical candle. A luminous intensity of ten-ninths of a spherical hefner is called an *international spherical candle*.

A standard candle and a Hefner lamp are shown in Fig. 57.

27. Lamps used as Secondary or Working Standards. — On account of flickering and low intensity, the Hefner amyl acetate lamp and the standard candle can be used only with great care. For the measurement of the luminous intensity of flames, the 10-c.p. Harcourt pentane lamp has been adopted by the London Gas Referees, and is in extensive use. In this lamp, *PL*, Fig. 106, the fuel is pentane vapor. The pentane flame is of about the color of illuminating gas flames and is much more steady than either the Hefner flame or the candle flame. The lamp, however, has the disadvantages of bulk, complicated construction, initial cost, and expense of maintenance. As it cannot be constructed to give an exact predetermined illumination, it must be calibrated by comparison with a Hefner lamp or standard candle.

After an incandescent lamp has been "aged" by being operated for some hours at an excess voltage, it will maintain for many hours a constant luminous intensity when operated at a constant voltage. By changing the impressed voltage, thereby changing the color of the light emitted, an incandescent working standard can be used for the photometric comparison of light sources of a considerable range of color. In calibrating an aged incandescent lamp for use as a secondary standard, a curve is constructed coördinating either current or impressed voltage, and luminous intensity expressed in either hefners or candle power.

28. Photometry. — The art of comparing luminous intensities is called photometry. As light-flux cannot be readily measured, luminous intensities cannot be directly measured by means of the defining equation (117).

If some standard of comparison be adopted for illumination, the relation expressed in (119) can be used for the determination of intensities. Thus, consider two point sources emitting light in all directions. Let a screen be placed between the two sources and normal to the line between them. If the intensities of the sources be I_1 and I_2, and the distances of the sources from the screen be r_1 and r_2, respectively, then the illuminations on the two sides of the screen are

$$E_1 = \frac{I_1}{r_1^2} \quad \text{and} \quad E_2 = \frac{I_2}{r_2^2}.$$

If the screen be placed so that the two sides are equally illumined, that is, until $E_1 = E_2$, then,

$$\frac{I_1}{r_1^2} = \frac{I_2}{r_2^2}.$$

Whence, $$\frac{I_1}{I_2} = \frac{r_1^2}{r_2^2}.$$ (120)

That is, when two luminous point sources equally illumine a screen normal to the line joining them, the luminous intensities of the sources are *directly* proportional to the squares of their distances from the screen. Though applying rigidly only to point sources, this relation can be used to a close degree of approximation for sources so small compared with their distances from the screen that, at the screen, the wave-fronts are nearly spherical.

29. The Equality of Brightness Method. — To compare luminous intensities by the principle developed in the foregoing article, a method is required for determining when two illuminations are equal. This necessitates the adoption of a criterion of equality of illumination. Two distinct criteria are in vogue.

Now light produces on the retina not only the sensation of color but also that of brightness. By means of the eye we can form a judgment whether or not two adjacent light spots of the same color have the same brightness. By moving relative to a screen two sources emitting light of the same color, we can adjust the brightness of two adjacent spots on the screen till to the eye they are the same. And if, after altering the position of one source, it be moved till the two spots again appear to be equally bright, it is found that the sources are in the same positions that they occupied when equal brightness was before observed. Thus we can assume that two illuminations of the same color are equal when they produce the sensation of equal brightness. This is the basis of the equality of brightness method for comparing illuminations.

In using this method for comparing luminous intensities, the sources are moved relative to a screen till the light spots produced by them on the screen appear to be equally bright. Then, from (120), the ratio of the intensities of the two sources equals the ratio of the squares of the distances of the screen to the two sources.

The equality of brightness method is satisfactory only when the two sources are approximately of the same color.

30. The Flicker Method. — It is a familiar fact that the sensation produced by light persists for a fraction of a second after the light has ceased to be incident on the retina. The duration of the sensation depends upon the illumination, but for ordinary illuminations is about 0.1 sec. For example, if a beam of light of

Fig. 58.

ordinary illumination entering the eye be interrupted oftener than about ten times per second there is no sensation of flickering.

If a steady stream of light from two sources enters the eye along the same path, the sensation due to the two will be fused into one. If the light from each source be interrupted regularly a few times per second, and pulses from one source be caused to alternate with

Fig. 59.

those from the other on the retina, the sensation will rise and fall, producing a flicker. If the streams be of unequal illumination, the sensation will rise and fall something as diagramed in Fig. 58. The speed of interruption can be increased till there is no flicker. But if the illuminations of the two beams are equal, a much lower speed of interruption will suffice to produce a steady sensation. The presence or absence of flicker at low rates of interruption is so sharp that it is taken as a criterion of equality of illumination. Thus, two beams of light which when interrupted alternately and incident on the same spot produce a steady luminous field, but which when either is altered in illumination produce a flickering field, are said to be of equal illumination.

In one form of apparatus for comparing luminous intensities by the flicker method due to Bechstein, Fig. 59, light from the two sources S_1 and S_2, Fig. 60, after being reflected by the diffusing wedge W, traverses the compound glass prism P_1P_2 and observing telescope T. The double prism consists of a prism P_1 fitted into a circular hole in another prism P_2 having the edge pointing in the opposite direction. The telescope is focalized on the faces of the diffusing wedge.

When the compound prism is in the position shown in the figure, light from S_1 that traverses P_1 is bent to the left out of the field of view, while the light that traverses P_2 emerges from the eye lens of the telescope. Light from S_2 traverses P_1 and the eye lens, while the light from S_2 that traverses P_2 is bent out of the field of

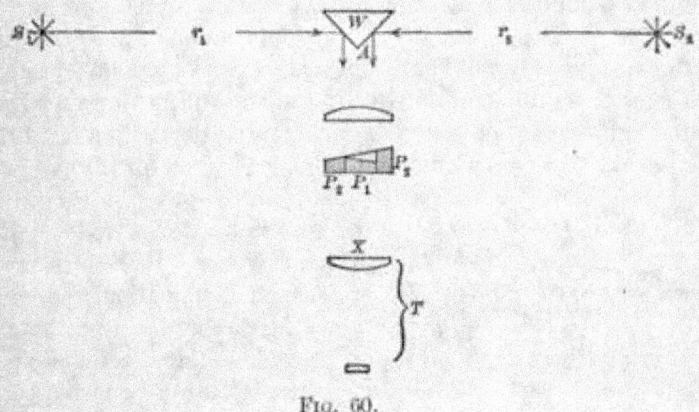

Fig. 60.

view. On looking through the eye lens one sees a bright circular disk due to light from S_2 surrounded by a bright ring due to light from S_1.

If the compound prism be turned 180° about the axis AX, the central bright disk will be due to light from S_1 and the circular ring to light from S_2. By rotating the compound prism by a motor, the disk and the ring are lighted alternately by S_1 and S_2. In general the field will flicker. But by moving the photometer, consisting of the diffusing wedge, compound prism and telescope, along the line joining the two light sources, a position will be found

at which the flicker disappears. At this position the two faces of
the wedge are equally illumined, and the ratio of the luminous
intensities of the two sources can be obtained by (120).

When the motor is at rest, the flicker photometer can be used as
an equality of brightness photometer.

31. Photometer Screens. — There are many devices to aid in
deciding when two spots of light are equally bright.

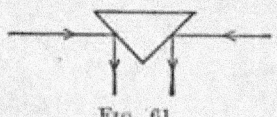

Fig. 61.

(a) The Ritchie wedge, Fig. 61, is a
prism of plaster-paris or other diffusing
material placed symmetrically with respect to the line joining the two light
sources. On looking toward the two
illumined faces, equality of brightness can be obtained by moving
the wedge along the line joining the two sources.

(b) The Bunsen screen consists of a grease spot on a piece of
white paper. When the paper is moved back and forward between two sources, the grease spot will in general appear bright on
one side and dark on the other. When
equally illumined on the two sides, however, the spot will disappear. In order
to see both sides at the same time mirrors
are placed as in Fig. 62.

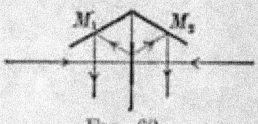

Fig. 62.

(c) The Leeson disk is similar to the Bunsen screen. It consists of a piece of white cardboard perforated by a star-shaped hole
and each face covered with a piece of white tissue paper. It is used
with mirrors like the Bunsen screen.

(d) The Joly cube consists of two rectangular blocks of paraffin
separated by a piece of tin foil. When placed between two light
sources with the tin foil normal to the line joining them, the two
halves of the cube will, in general, appear unequally bright. At
one position, however, the two halves are equally bright and the
division line between the two halves disappears.

(e) The "Photoped" used by the London Gas Referees is com-
posed of a piece of white paper at one end of a metal tube, Fig. 63,
two adjacent sections of which are illumined by the two sources
under comparison. Light from S_1 traversing the rectangular
aperture K illumines the section AB, and light from S_2 illumines

the section BC. The length h of the tube is adjusted so that the
illumined areas just meet at B without overlapping. The distance
r_1 or r_2 is adjusted till the line of separation at B disappears.

   (f) The Lummer-Brodhun screen is very much used for the
comparison of the luminous intensities of sources of the same color.

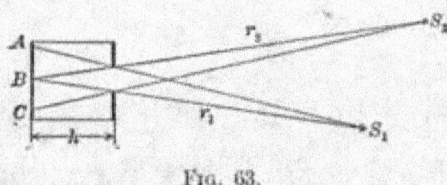

FIG. 63.

Light from the two sources after being scattered by the white
screen S is reflected by the mirrors M_1 and M_2 into a pair of total
reflecting prisms p_1 and p_2. A part of the hypotenuse face of
one prism is ground away as shown in
Fig. 64. Light incident on the contact face will be transmitted by both
prisms, whereas light incident on a part
of the hypotenuse face of either prism
not in contact with the other prism
will be totally reflected. Suppose a
ring has been ground off the hypotenuse face of p_1. Then there will emerge
from p_2 a cylinder of light from the
left source and a concentric tube of
light from the right source. The total
reflecting prism p_3 serves to direct
these beams into a telescope T placed
perpendicular to the line joining the

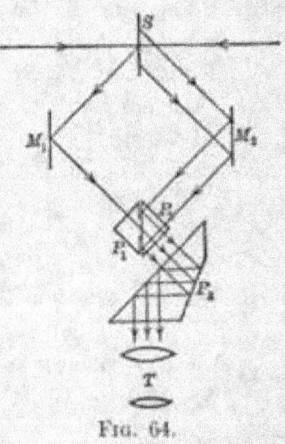

FIG. 64.

two sources. As no light crosses the face of any prism except
perpendicularly, there is no dispersion.

   32. The Cardinal Points of a Lens or Lens System. — A lens
or lens system that causes an incident plane wave to emerge as a
convergent wave is called a converging or positive lens or system.
The point toward which a plane wave advancing parallel to the

principal axis converges after emergence is called the *principal focus* of the lens or system. A lens or lens system that causes an incident plane wave to emerge as a divergent wave is called a diverging or negative lens or system. In the case of a negative lens or system, the principal focus is the point from which appears to diverge an emergent wave that on incidence was a plane wave advancing parallel to the principal axis.

There are two planes normal to the principal axis of a lens or lens system, called "principal planes," which possess the property that the prolongation of any incident ray meets the first, and the prolongation of the corresponding emergent ray meets the second, at points equally distant from the principal axis. The points where the principal planes are cut by the principal axis are called the *principal points* of the lens or system. There are a pair of principal points for light of each color.

The *principal focal length* of a lens or system is the distance between the principal focus and the corresponding principal point of emergence.

In the case of any lens or system there are two points, called *nodal points*, which have the property that if incident light be directed toward one, the emergent light will proceed in a parallel direction from the other. If parallel light be incident on a lens or system, the direction of the emergent light will be unaffected by a rotation of the lens about an axis through the nodal point of emergence perpendicular to the principal axis. This furnishes a convenient experimental method for locating nodal points.

The principal foci, principal points, and nodal points constitute a system called the "cardinal" points of the lens or system. Knowing the cardinal points of a given lens or system, the emergent ray corresponding to any assigned incident ray can be constructed. In the ordinary case in which the lens is bounded on both sides by the same medium, the principal points coincide with the nodal points. The term "equivalent points" is used to denote these superimposed nodal and principal points.

It follows that if parallel light be incident axially on a lens or system bounded on both sides by the same medium, and the point be found about which the lens or system can be slightly rotated

without changing the position of the focus, the principal focal
length is the distance between the axis of rotation and the principal
focus. The principal focal length of a single lens is different for
light of different colors as well as for different parts of the lens
traversed by the light. For a lens having large aberrations, the
principal focal length is taken as the distance from this axis of
rotation to the circle of least confusion (Art. 35).

33. Chromatic Aberration. — A pencil of light the axis of
which coincides with the principal axis of a lens is called a "direct
axial pencil." A wide pencil is
called a "beam."

When a pencil of parallel white
light is incident upon a simple
converging lens, light of different
frequencies is brought to different

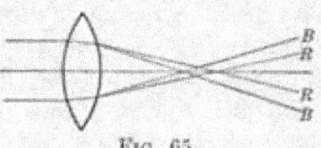

Fig. 65.

foci. The property of a lens in virtue of which parallel light
of different frequencies incident on the same part of a lens is not
refracted to a single focus is called *chromatic aberration*. Due to
chromatic aberration there will be as many images of an object
as there are component colors in the light. These images will be

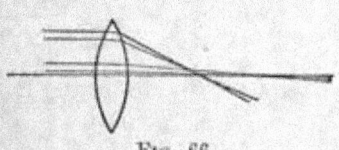

Fig. 66.

of different sizes and will be
at different positions — the blue
image being the smallest and
nearest the lens, and the red being farthest away and largest.
For example, with sunlight, if a
white screen be placed in the position of the red image, there will
be superimposed on the red image other images of different sizes
and colors that are not in focus. The result will be a nebulous
spot surrounded by a blue border.

34. Longitudinal Spherical Aberration. — Monochromatic light
advancing parallel to the principal axis and incident near the
margin of a convex lens bounded by spherical faces will be more
refracted than will a direct axial pencil (Fig. 66). Thus, the
focus for marginal pencils is nearer the lens than the focus for
centric pencils. The longitudinal distribution of the focus of an
axial beam is called *longitudinal* or *axial spherical aberration*.

35. Circle of Least Confusion. — Any simple lens has both chromatic and longitudinal spherical aberration. For such a lens there is no sharp focus for a beam of white light. In Fig. 67 are shown two axial pencils of white light incident near the margin of a convex lens, and also two incident near the center. On emergence,

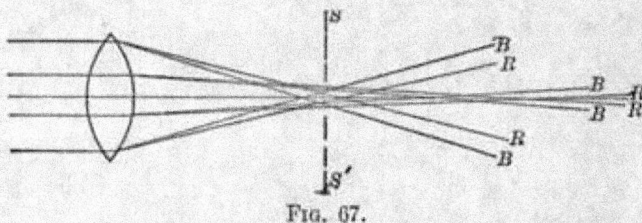

FIG. 67.

each pencil is divided into its component colors, of which only the red and the blue are indicated in the figure. If a screen SS' normal to the principal axis be placed in the narrow part of the emergent beam, one will observe a bright spot surrounded by a colored ring.

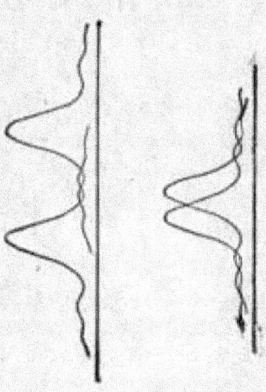

FIG. 68.   FIG. 69.

At a certain position the diameter of the spot will be a minimum. The bright spot of smallest diameter is called the "circle of least confusion."

In the case of a lens having large aberrations, the circle of least confusion is taken to be the focus for a wide beam of white light.

36. The Resolving Power of a Lens. — The image of a point source consists of a bright central diffraction disk surrounded by a series of concentric diffraction bands. The distribution of light in the images of two luminous points is as represented in Fig. 68. If the images of two points overlap as much as in Fig. 69, they cannot be distinguished as images of separate points. When two images can be distinguished as separate images, the two point sources are said to be "resolved." It is commonly assumed that the smallest distance between two point sources that can be resolved is that which will produce images whose centers are separated by a distance equal to the radius of the diffraction disk of one of them. When the images of two point sources can be just distinguished as separate, the lens is said to be at its "limit of resolution." The ability of a lens to separate the images of two adjacent point sources is called the "resolving power" of the lens. The "limiting angle of resolution" of a lens is the angle at the center of the lens between lines from two object points that are just resolved by the lens. The angular resolving power of a lens is the reciprocal of the limiting angle of resolution.

For a lens of small aperture compared with its focal length an approximate value of the resolving power can be readily obtained.

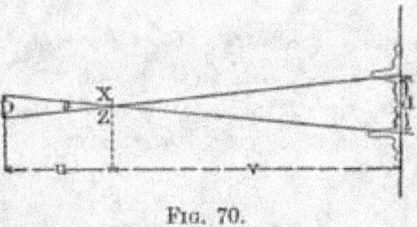

FIG. 70.

Consider the case of two point sources, Fig. 70, separated by a linear distance D. If the distance between the images be d, then from the geometry of the figure,

$$\frac{D}{u} = \frac{d}{v}.$$

Denoting by θ the angular distance at the lens between the two sources, we have, since θ is small,

$$\theta = \frac{D}{u} = \frac{d}{v}. \tag{121}$$

When the linear distance between the centers of the images is less than the radius of the central diffraction disk of one of them, the two images cannot be distinguished as two separate images. Whence, putting $d = r$ in the above equation, it follows that two distinct images of two point sources cannot be obtained if the angular distance at the lens, between the sources, is less than

$$\theta = \frac{D}{u} = \frac{r}{v}. \qquad (122)$$

From the second of these equations another value of the limiting angle of resolution can be obtained which involves no quantities except the aperture of the lens and the wave-length of the light transmitted. We shall first consider the lens to be covered with a diaphragm provided with a narrow slit of width a normal to the

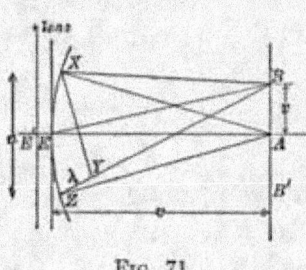

FIG. 71.

plane of the diagram, Fig. 71. The image of a point source will be found on the screen distant v from the lens. If the point source be on the principal axis, light, which after traversing the slit and lens arrives at A, will at that point be in the same phase, that is, the center of the image will be at A. At points B, B', on either side of A and at

such a distance from A that the difference between ZB and XB (or XB' and ZB') is one wave-length λ, there will be destructive interference. The approximate half width ($r = AB$) of this diffraction band is now to be found and substituted in (122).

Lay off $BY = BX$, and draw XY and EB. Since ZY, which equals one wave-length, is minutely small compared with the distance ZB, the line XY is almost perpendicular to ZB, and consequently the triangles XYZ and EAB are approximately similar. Then

$$\frac{AB}{EB} = \frac{ZY}{ZX}.$$

At the limit of resolution AB is so small compared with EA that $EA = EB$. Thus we may write the above equation in the form

$$\frac{AB}{EA} = \frac{ZY}{ZX},$$

or, using the notation in the figure,

$$\frac{r}{v} = \frac{\lambda}{a},$$

or
$$r = \frac{\lambda v}{a}. \tag{123}$$

On substituting in (122) this value of r, we obtain for the limiting angle of resolution of a narrow diametral stripe of width a,

$$\theta_s = \frac{\lambda}{a}, \tag{124}$$

and for the angular resolving power of the stripe of the lens,

$$\left(\frac{1}{\theta}\right)_s = \frac{a}{\lambda}, \tag{125}$$

where θ is expressed in radians.

When the aperture is a centric circle instead of a diametral slit, the place of maximum interference is somewhat farther from the lens than in the case just considered, and the radius of the diffraction disk is about 1.22 times that given in (123). Thus, when light traverses an unstopped lens or one provided with a diaphragm containing a centric circular aperture of diameter a,

$$r = \frac{1.22\,\lambda v}{a}, \tag{126}$$

the limiting angle of resolution is

$$\theta_c = \frac{1.22\,\lambda}{a}, \tag{127}$$

and the angular resolving power is

$$\left(\frac{1}{\theta}\right)_c = \frac{a}{1.22\,\lambda}. \tag{128}$$

37. The Smallest Resolvable Detail in an Image. — It will now be shown that the smallest resolvable detail in an image is proportional to the ratio of the focal length of the lens to its aperture.

From (122) and (126), the value of the least angular separation between two object points is

$$\theta = \frac{r}{v} = \frac{1.22\,\lambda}{a},$$

where a is the diameter of the round centric aperture of the lens, v is the image distance, λ is the wave-length of the light, and r is the distance between two image points at the limit of resolution. In this equation r is taken to be the radius of the diffraction disk of one of the image points.

Writing the above equation in the form

$$\frac{r}{\lambda} = \frac{1.22\,v}{a}, \tag{129}$$

we see that the smallest resolvable detail in the image, expressed in wave-lengths, is proportional to the ratio of the focal length of the lens to its aperture.

Since for a system of lenses the effective aperture is that of the smallest aperture in the system, and since λ is the same for each lens, it follows that the resolving power of a lens system is that of the lens which has the smallest resolving power.

38. Absolute and Relative Indices of Refraction. — The ratio of the speeds of light in vacuo to the speed in a given medium is called the absolute index of refraction of the medium. The ratio of the speeds of light in any two media is called the index of refraction of the second medium relative to the first. Thus, representing the speeds of light of a given frequency in vacuo, and in two media by s, s_1, and s_2, respectively, then the absolute indices of refraction of the two media are

$$n_1 = \frac{s}{s_1} \quad \text{and} \quad n_2 = \frac{s}{s_2},$$

and the index of refraction of the second medium relative to the first is

$$n_{2,1} = \frac{s_1}{s_2} \tag{130}$$

$$= \frac{n_2}{n_1}. \tag{131}$$

It is readily shown that when light goes from one medium to another in which the velocity is different, the ratio of the sines of the angles between the normal to the interface and the rays in the two media equals the ratio of the speeds in the two media.   That is,

$$\frac{\sin \phi_1}{\sin \phi_2} = \frac{s_1}{s_2}.$$

Consequently, (131),

$$n_{2,1} \left[= \frac{n_2}{n_1} \right] = \frac{\sin \phi_1}{\sin \phi_2}. \tag{132}$$

This is called Snell's law.

39. The Refractive Index of a Substance in the Form of a Prism. — Snell's law is the basis of most methods for the determination of refractive indices.   The above expression of the law is often written

$$n = \frac{\sin i}{\sin r}, \tag{133}$$

where n is the index of refraction of the second medium relative to the first, i is the angle of incidence and r is the angle of refraction.
Starting with this equation, a formula will now be deduced for the determination of the refractive index of a substance in the form of a prism of known refracting angle.

In Fig. 72, consider light incident on the prism of refracting angle A at an angle of incidence i. On entering the prism, the angle of refraction is r.

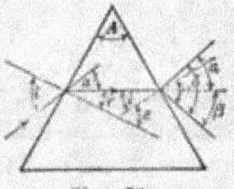

FIG. 72.

At the second face, the angle of incidence is represented by i', and on emergence into the air the angle of refraction is r'.   At the point of emergence draw a line parallel to the ray incident on the first face.   The angle δ between this line and the emergent ray is the deviation produced by the prism.

From the geometry of the figure, the deviation

$$\delta = \alpha + \beta.$$

But

$$\alpha \left[= \alpha' \right] = i - r \quad \text{and} \quad \beta = r' - i',$$

so that

$$\delta = i - r + r' - i'. \tag{134}$$

But

$$r + i' = \epsilon = A. \tag{135}$$

Consequently,

$$\delta = i + r' - A. \tag{136}$$

On arranging the experiment so that $r' = i$, $r = i'$. In this case we have from (135), $r = \frac{1}{2} A$ and from (136), $\delta = 2i - A$ or $i = \frac{1}{2}(A + \delta)$.

On substituting these values of i and r in (133), we have, under the conditions of the experiment,

$$n \left[= \frac{\sin i}{\sin r} \right] = \frac{\sin \frac{1}{2}(A + \delta)}{\sin \frac{1}{2} A}. \tag{137}$$

In the following article it is shown that under the condition implied in (137), namely, that the angle of emergence r' equals the angle of incidence i, the deviation δ is the smallest that can be produced by the given prism. A prism in this position is said to be set for minimum deviation.

40. The Condition that the Deviation produced by a Prism shall be Minimum. — From (134), the deviation

$$\delta = (i + r') - (i' + r). \tag{138}$$

If the medium surrounding the prism be air or anything else having a smaller refractive index than the prism,

$$(i + r') > (i' + r).$$

We shall now find under what conditions the value of $(i + r')$ is the smallest possible. From Fig. 72 and Snell's law,

$$\sin i = n \sin r \quad \text{and} \quad \sin r' = n \sin i'.$$

Whence, $\sin i - \sin r' = n(\sin r - \sin i')$

and $\sin i + \sin r' = n(\sin r + \sin i')$.

Expanding each of these expressions we obtain

$$2 \sin \tfrac{1}{2}(i - r') \cos \tfrac{1}{2}(i + r') = 2n[\sin \tfrac{1}{2}(r - i') \cos \tfrac{1}{2}(r + i')],$$
$$2 \sin \tfrac{1}{2}(i + r') \cos \tfrac{1}{2}(i - r') = 2n[\sin \tfrac{1}{2}(r + i') \cos \tfrac{1}{2}(r - i')]. \tag{139}$$

Dividing each member of the last equation by the corresponding member of the preceding, we obtain

$$\tan \tfrac{1}{2}(i + r') \cot \tfrac{1}{2}(i - r') = \tan \tfrac{1}{2}(r + i') \cot \tfrac{1}{2}(r - i')$$

or $$\tan \tfrac{1}{2}(i + r') \tan \tfrac{1}{2}(r - i') = \tan \tfrac{1}{2}(r + i') \tan \tfrac{1}{2}(i - r').$$

Whenever $(i + r') > (i' + r)$, then $\tan \tfrac{1}{2}(i + r') > \tan \tfrac{1}{2}(i' + r)$, and we see from the preceding equation that

$$\tan \tfrac{1}{2}(i - r') \lessgtr \tan \tfrac{1}{2}(r - i'),$$

or $$(i - r') \lessgtr (r - i')$$

and $$\cos \tfrac{1}{2}(i - r') \gtrless \cos \tfrac{1}{2}(r - i').$$

This result in connection with (139) shows that

$$\sin \tfrac{1}{2}(i + r') \lessgtr n \sin \tfrac{1}{2}(r + i').$$

Whence, the smallest value of $(i + r')$ is such that

$$\sin \tfrac{1}{2}(i + r') = n \sin \tfrac{1}{2}(r + i'),$$

that is, (139), when

$$\cos \tfrac{1}{2}(i - r') = \cos \tfrac{1}{2}(r - i'),$$

or $$i - r' = r - i'.$$

Putting this equation into the form

$$i - r = r' - i',$$

we see that minimum deviation occurs when the bending at the first face of the prism equals that at the second face. But in order that there shall be the same bending, the angles between the rays in either medium and the corresponding normals to the interfaces must be equal. That is, for minimum deviation

$$i = r' \qquad \text{and} \qquad r = i'.$$

41. The Critical Angle of Incidence. — An important special case of (132) is one in which light goes from a medium in which the speed is less to a medium in which the speed is greater. In this case, $s_2 > s_1$ and $\phi_2 > \phi_1$. In the limit when $\phi_2 = 90°$, we have from (131) and (132),

$$\frac{n_1}{n_2} \left[= n_{1,2} \right] = \frac{1}{\sin \phi_1},$$

or, in words, when light goes from a medium in which the speed is
less to one in which the speed is greater, and at such an angle of
incidence that the angle of refraction equals 90°, then the index of
refraction of the first medium relative to the second equals the
reciprocal of the sine of the angle of incidence. If the angle of
incidence be smaller than this critical value, some light will be
transmitted into the second medium and the remainder will be
reflected back into the first medium. If the angle of incidence
be greater than this critical value, no light will be transmitted into
the second medium but all will be reflected back into the first
medium. Denoting the critical angle of incidence by c, we can
write the above equation in the form

$$\frac{n_1}{n_2} = n_{1,2} = \frac{1}{\sin c}. \tag{140}$$

From this relation will now be derived the equation upon which
depend several important instruments for the determination of

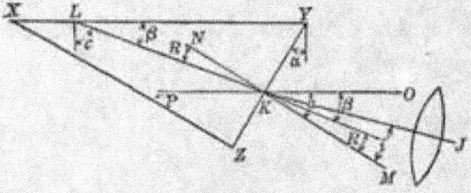

Fig. 73.

refractive indices of liquids. Suppose the liquid whose refractive
index is sought is on the face XY of the glass block XYZ, Fig. 73.
Consider the ray KL which makes with the normal to XY the
critical angle c. In so far as the following analysis is concerned,
the light may travel either in the direction KL or LK. Denoting
the refractive index of the specimen by n, and that of the prism by
n_g, we may write, (140),

$$\frac{n_s}{n_g} = \sin c.$$

Through the point K draw the line NM perpendicular to YZ
and the line OP parallel to XY. Produce LK. Let KJ be the
continuation, in air, of the ray LK.

Represent the complement of the angle XYZ by α; the angle YLK by β; JKM by i; and LKN by R. Then, from the geometry of the figure,

$$\sin c = \cos \beta = \cos (\alpha - R).$$

Consequently, the preceding equation may be written

$$
\begin{aligned}
n_e &= n_g \cos (\alpha - R) \\
&= n_g (\cos \alpha \cos R + \sin \alpha \sin R) \\
&= n_g (\cos \alpha \sqrt{1 - \sin^2 R} + \sin \alpha \sin R).
\end{aligned}
$$

If the index of refraction of glass be taken relative to air, we have from (132)

$$\frac{n_g}{1} = \frac{\sin i}{\sin R}, \quad \text{or} \quad \sin R = \frac{\sin i}{n_g}.$$

Substituting this value of $\sin R$ in the preceding equation,

$$
\begin{aligned}
n_e &= n_g \cos \alpha \sqrt{1 - \frac{\sin^2 i}{n_g^2}} + \sin \alpha \frac{\sin i}{n_g} \Big) \\
&= \cos \alpha \sqrt{n_g^2 - \sin^2 i} + \sin \alpha \sin i.
\end{aligned}
\qquad (141)
$$

In the Pulfrich refractometer, the angle α is $0°$. In this case, $\sin \alpha = 0$, $\cos \alpha = 1$, and (141) becomes

$$n_e = \sqrt{n_g^2 - \sin^2 i}. \qquad (142)$$

In the Abbé, the butyro, and the Zeiss immersion refractometer, the angle α has different values depending upon the indices of refraction the particular instrument is designed to measure.

A plane wave, whose front is perpendicular to the plane of the paper on entering the liquid-glass surface XY at grazing incidence, will emerge from the glass into the air with the wave front perpendicular to the plane of the paper and making an angle i with the normal to the glass-air surface. If, after emergence, this light traverses axially a converging lens, it will be brought to a point focus in the plane of the paper. The line through this point and perpendicular to the plane of the paper will contain the foci of all beams, which, on entering the liquid-glass surface at grazing incidence, are not parallel to the plane of the paper.

Light incident on the liquid-glass surface at angles less than 90°
will emerge from the glass into the air below the line *KJ*. Consequently, if monochromatic light be used, the field of view in the
lens consists of a bright part and a dark part sharply separated
from one another by the line *KJ*. If white light be used, the
division line will be nebulous and colored.

42. Specific Refractivity. — The refractive index of a substance depends upon the temperature, pressure, and state of
aggregation of the substance. But from a consideration of the
Clausius-Mossotti theory of dielectrics and the electromagnetic
theory of light, Lorenz of Copenhagen and Lorentz of Leyden have
found a function of the refractive index and density that is independent of the above-mentioned variables. This is of the form

$$\frac{N^2 - 1}{N^2 + 2} \cdot \frac{1}{d},$$

where N is the refractive index for light of very long wave-length,
and d is the density. Experiment shows that this function is not
only practically independent of temperature, pressure, and state
of aggregation, but also of the wave-length of light employed in
determining the refractive index. It is called the Lorenz specific
refractivity or specific refracting power, and may be expressed in
the form

$$R_L = \frac{n^2 - 1}{n^2 + 2} \cdot \frac{1}{d}, \tag{143}$$

where n is the refractive index for light of any wave-length.

A simpler function of refractive index and density is found by
experiment to be nearly independent of temperature and pressure,
though not of state of aggregation. This is called the Gladstone
and Dale specific refractivity or specific refracting power, and is of
the form

$$R_{GD} = \frac{n - 1}{d}, \tag{144}$$

where n is the refractive index for light of a given wave-length
and d is the density.

The specific refractivity of a substance is practically unin-fluenced by the presence of another substance that may be mixed with it. Thus, if p per cent of one substance of specific refrac-tivity R_1 be mixed with $(100 - p)$ per cent of another substance of specific refractivity R_2, the specific refractivity of the mixture will be

$$R = R_1 \frac{p}{100} + R_2 \frac{100 - p}{100}, \qquad (145)$$

where the R's may be either Lorenz's or Gladstone and Dale's specific refractivities. This equation is useful for the determina-tion of the specific refractivity of one ingredient when one knows the composition of the mixture together with the specific re-fractivity of the mixture and of the other ingredient. In the case of a mixture of two ingredients, knowing the specific refractivity of the mixture and of each ingredient, (145) can be used to deter-mine the composition.

The product of the specific refractivity of a substance and its molecular weight is called the *molecular refractivity* of the sub-stance. There is the Lorenz molecular refractivity and also the Gladstone and Dale molecular refractivity.

43. The Pulfrich Refractometer. — A horizontal beam of monochromatic light from a Geissler tube G, Fig. 74, after being converged by a condenser C on to the horizontal liquid-glass sur-face XY of the glass block P, Figs. 74 and 75, emerges from the vertical glass-air surface, and by means of a 45° prism in front of the objective of a telescope T is reflected along the axis of the telescope.

The field of view of the telescope consists of a bright region and a dark region separated by a sharp line as explained at the end of Art. 41. Light coming to the boundary line makes an angle i with the normal to the vertical face of the rectangular block P.

The angle i may be measured by first rotating the divided disk D with the attached telescope T about the axis through A, till the boundary line is brought into coincidence with the cross hairs in the focal plane of the telescope; noting the scale reading; then rotating the divided disk till the cross hairs coincide with a ray

normal to the vertical face of P; and again noting the scale reading.   The difference between these scale readings is the angle i.
   The coincidence of the cross hairs with a normal from the vertical

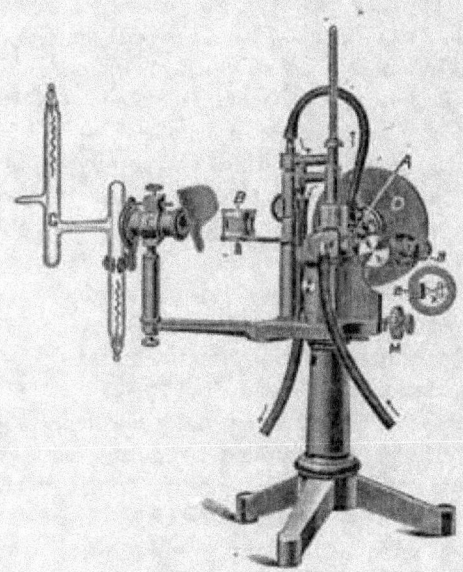

Fig. 74.

face of the block P is obtained by the aid of a tiny reflecting prism
a in contact with the cross hairs.   Light entering this prism
through a hole in the side of the telescope is
reflected past the cross hairs to the vertical
face of P.   It is there reflected back through
the objective.   When the image of the cross
hairs coincides with the cross hairs, as shown at
x in the small diagram in the right-hand margin
of Fig. 74, then the normal to the vertical face

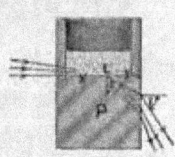

Fig. 75.

of P passes through the intersection of the cross hairs.   The scale
reading now observed is the zero point for subsequent observations.
   By means of an accessory reflecting prism and attached lens B,
light from a sodium flame can be quickly substituted for the light
from the Geissler tube.

The Pulfrich block P and the liquid under test can be maintained at a definite temperature by means of a constant stream of water through the tubes shown in Fig. 74.

For increasing the range of the instrument, glass blocks of different refractive indices are supplied. For example, one block can be used to measure refractive indices from that of water, (1.33), up to 1.61; another has a range of from 1.47 to 1.74.

By dividing the specimen cell into two compartments by means of a thin black glass partition in the plane of the incident light, the difference between the refractive indices of two liquids can be determined from a single observation. With this arrangement there will be two critical border lines and the angle between these lines is a measure of the difference between the refractive indices of the two specimens.

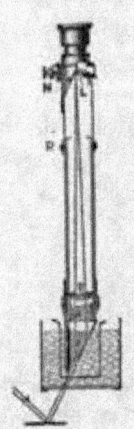

The scale of the instrument is usually divided into degrees. Tables are supplied by means of which the refractive index corresponding to any scale reading when a particular glass block was employed can be obtained without computation.

With this instrument, readings of refractive indices accurate to one unit in the fourth decimal place are possible; and when using the double cell, readings of differences of refractive indices can be made with a precision of one unit in the fifth decimal place.

FIG. 76.

44. The Zeiss Immersion or Dipping Refractometer. — This instrument consists of a telescope having a glass micrometer eyepiece, an Amici prism A, Fig. 76, and a glass block P which performs the same function as the rectangular block in the Pulfrich refractometer. As in the latter instrument, light meeting at grazing incidence the surface separating the glass block P and the liquid under test is refracted at the critical angle. In this instrument, however, the angle α has such a value that the light transmitted by the glass block will proceed along the axis of the telescope. The position of the critical line separating the dark from the light portion of the field of view is marked by the eyepiece micrometer. According to the type of ocular, this scale may be engraved on one surface of the field lens F, or on a piece of plain glass situated either between the objective and field lens or between the field lens and eye-lens. By means of the micrometer screw M, the glass scale can be moved laterally, thereby permitting an accurate determination of a fraction of a scale division.

The liquid under test is contained in a glass beaker in a constant temperature bath provided with a glass bottom. A mirror reflects daylight or lamplight at grazing incidence to the surface separating P from the liquid under test. This light will be dispersed in planes parallel to the paper by the specimen and the glass block P. If the dispersed light be not recombined, the critical line separating the bright from the dark part of the field of view will be colored and nebulous. The function of the Amici prism A is to cause the light dispersed perpendicular to the critical line to recombine into white light. It consists of a flint and two crown glass prisms the same as used in the direct vision spectroscope, Fig. 89. Such a prism will disperse incident white light in the plane of the triangular faces without changing the direction of the light of the particular frequency for which the prism is designed. Conversely, when a beam of light dispersed in the plane of the triangular faces is transmitted by such a prism, it will suffer a further dispersion when the prism is in one position, and a diminution of dispersion when the prism is rotated about the axis of the beam. In this instrument, the Amici prism acts as a compensator of the dispersion produced by the specimen and the block P. The amount of compensation perpendicular to the critical line depends upon the orientation of the compensator relative to the block P. This orientation is effected by twisting the ring R.

The scale of this instrument is divided into 100 parts. The refractive indices corresponding to the various divisions have been computed and tabulated in a convenient form supplied with the instrument. The range of the instrument is from 1.3254 to 1.3664. Settings can be made with a precision of four units in the fifth decimal place.

45. The Abbé Refractometer. — The action of this instrument involves the same principles as does the action of the Zeiss immersion refractometer. The specimen is in the form of a thin stratum between a triangular glass block P, Fig. 77, similar to the one used in the Zeiss immersion refractometer, and a supplementary glass block P'. (In the figure, these two blocks are shown separated, ready for the insertion of a few drops of the specimen.) The face of the supplementary block P' in contact with the specimen is ground. Ordinary light reflected up through this block renders each point of the ground face a light source. Light from these point light sources, meeting the surface separating the specimen from the block P at grazing incidence, is refracted at the critical angle, and after traversing the telescope objective is converged to a critical line as in the Pulfrich and the Zeiss immersion refractometers.

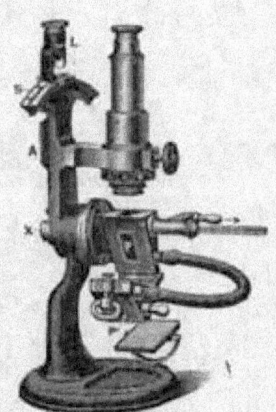

The dispersion produced by the specimen and the block P is compensated by means of two Amici prisms, coaxial with the telescope and situated between the block P and the telescope objective. By means of a milled head, these

FIG. 77.

Amici prisms can be simultaneously rotated in opposite directions till a position is found for which the critical line is colorless and sharp.

In making a setting with a given specimen, the pair of glass blocks P and P' is rotated as a unit about a horizontal axis through X, by means of a pointer or alidade A, till the critical line separating the light from the dark portion of the field of view coincides with a fixed cross hair in the eyepiece of the telescope. The reading on the fixed circular scale S is facilitated by the use of a lens L attached to the end of the alidade. Usually, this scale is divided to give refractive indices directly, without computation, for sodium light.

The angle between the planes of incidence of the two Amici prisms of the compensator is indicated by a circular scale on a drum around the objective. When the critical line is colorless, this angle together with the constants of the instrument supply the data for computing the mean dispersion produced by the specimen.

The range of this instrument is usually from 1.3 to 1.7, and the precision is about two units in the fourth decimal place.

46. The Butyro Refractometer. — In many lines of industrial work it is required to make frequent tests of the integrity or purity of specimens of the same substance. When a number of specimens are presumably of the same material, and consequently should have the same refractive index and the same mean dispersion, the test can often be made with a much simplified form of the Abbé refractometer. The two-Amici-prism compensator can be omitted. The glass block P can be rigidly attached to the objective end of the telescope and the alidade with the circular scale dispensed with.

To produce the necessary compensation for one particular substance, the refracting angle of the block P is given such a value that with the material for which the instrument is designed, the critical line will be white, and will fall on a determined division line of a glass scale in the eyepiece of the telescope.

If a specimen be tested which has a refractive index different than that for which the instrument was designed, the position of the critical line will indicate the fact. If the specimen has a mean dispersion different than that for which the instrument was designed, the critical line will be colored. In case the mean dispersion of the specimen is less than that for which the instrument was designed, the red component of the transmitted light will be deviated to the bright side of the critical line, and the blue to the dark side. Thus, the critical line appears red. With a specimen having a mean dispersion greater than that which the instrument achromatizes, the blue light will be deviated to the bright side of the critical line and the red to the dark side. In this case the critical line will appear blue.

An instrument of this sort designed for testing butter-fat is called a butyro refractometer. Refractometers of the same sort are used for testing various other commercial products. They have a precision of about one unit in the fourth decimal place.

47. The Féry Refractometer. — This instrument, Figs. 78 and 79, consists of a collimator C and telescope T, between which is an acute-angled hollow glass prism P inside of a water bath provided with ends formed of lenses. The water bath is maintained at a

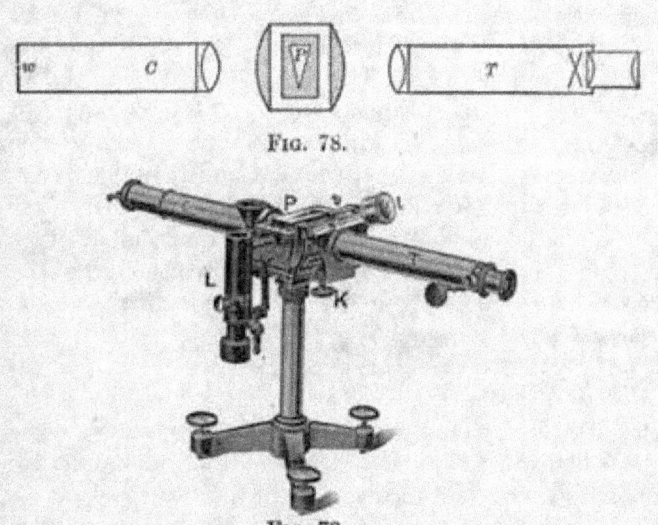

FIG. 78.

FIG. 79.

constant known temperature by means of a convection current produced by a lamp L. The water bath with the contained prism can be moved perpendicularly to the common axis of the collimator and telescope by means of a rack-and-pinion operated by the knob K. A setting consists in moving as a unit the water bath with the prism containing the

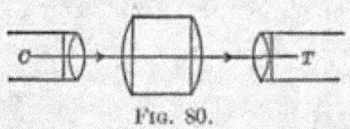

FIG. 80.

liquid whose refractive index is required, until the image of a wire w coincides with the cross hairs in the eyepiece. The amount of the displacement from the zero position is read by means of a scale s and vernier v.

If the prism were not in the water bath, Fig. 80, a beam of light from the collimator, entering the water bath axially, would emerge from the water bath in the same direction and form an image of the wire w on the cross hairs of the telescope. On introducing the prism containing the specimen, the beam will be refracted toward the base of the prism, Fig. 81. But by moving as a unit the water bath and the contained prism, in a direction transverse to the common optic axis of the collimator and telescope, Fig. 82, the light emerging from the water bath will be again axial. It will now be shown that the linear distance the

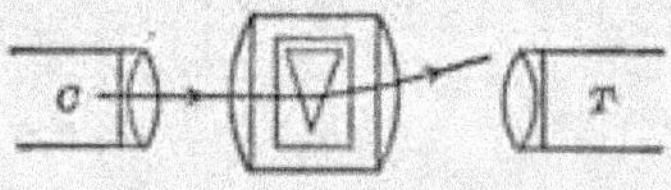

Fig. 81. 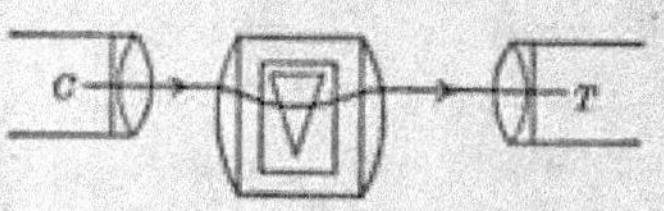   Fig. 82.

water bath must be moved to neutralize the deviation produced by the prism of specimen is directly proportional to the refractive index of the specimen.

It is shown in Arts. 39 and 40, that for a prism of refracting angle A set at minimum deviation, the index of refraction of the material of which the prism is made

$$n = \frac{\sin \frac{1}{2}(A + \delta)}{\sin \frac{1}{2}A},$$

where δ is the angle of deviation.

If the refracting angle of the prism be sufficiently acute, the angles A and δ will be so small that the sines, and the angles (expressed in radians) will be approximately equal. In this case, the preceding equation becomes

$$n = \frac{A + \delta}{A},$$

or

$$\delta = A(n - 1).$$

Whence, for an acute-angled prism, the index of refraction is directly proportional to the devia ion.

Now in any plane through the axis, the lenses forming the end faces of the water bath are essentially prisms of variable angle, the angle varying directly with the distance from the axis. It follows that by moving the water bath transversely, the deviation produced by the liquid prism will be neutralized by an amount directly proportional to the linear displacement of the water bath from the axial position. Consequently, when the deviation of the prism is exactly neutralized, the linear displacement of the movable system from the zero position is directly proportional to the refractive index of the specimen.

The scale of the commercial instrument is divided so as to express refractive indices directly. The numbers on the scale represent the first two decimal places of the refractive indices — the whole number unity being always understood to precede the reading. By means of the vernier divided into 25 parts one can read the third and fourth decimal places.

Before the specimen is introduced into the hollow prism, the instrument is adjusted as follows: Fill the water bath, light the lamp, and wait till the bath is at about the required temperature. By means of the milled head attached to the jaws of the collimator slit, bring into coincidence the two "0" marks on the jaws. Set the vernier to read the refractive index of the glass composing the hollow prism. Place a sodium flame in front of the slit. Move the eyepiece in and out till the cross hairs are in focus. By means of the large milled head on the telescope, move in and out the fitting containing the eyepiece and the cross hairs till the image of the wire w is in the plane of the cross hairs. By means of the small milled head on the telescope move the cross wires transversely till their intersection coincides with the image of the wire w. The instrument is now in adjustment for refractive indices between 1.3300 and 1.5326.

In making a determination, the specimen is introduced into the hollow prism and the milled head K turned till the image of the wire w is again on the intersection of the cross hairs. The refractive index of the specimen is then read directly by means of the scale and vernier.

The range of the instrument can be altered by changing the thickness of the specimen traversed by the light. The standard instrument can be adjusted as follows so that the scale indications are just 0.14 too small. When it is desired to increase the range by this amount, the jaws of the collimator are moved till the two

"1" marks coincide; the vernier is set to read 0.14 less than the refractive index of the glass composing the hollow prism; and the remainder of the adjustments made as above described.

48. The Oleorefractometer of Amagat and Jean. — Like the Féry refractometer, this instrument consists of a collimator, telescope and hollow glass prism. It differs in that the specimen prism is within a parallel-sided cell filled with a standard liquid; this cell is enclosed in a parallel-sided water bath; and the prism, cell, and water bath are fixed in position relative to the collimator and telescope. By means of an absorptive glass in the collimator, light from a lamp is rendered sufficiently monochromatic.

The position of the collimator window is such that the field of view in the telescope consists of a bright portion and a dark portion separated by a vertical straight line. If the refractive index of the specimen be different from that of the standard liquid, the light will be deviated either to the right or to the left, according as the refractive index is greater or less than that of the standard liquid. Thus, the position of the critical line depends upon the difference between the refractive indices of the two substances. By always using the same liquid in the cell, the refractive index of the specimen in the prism will be indicated by the position of the critical line. This position is read on a uniformly divided eyepiece micrometer (Art. 12) within the telescope. Before introducing the specimen, the prism and cell are both filled with the standard liquid, and the position of the collimator window adjusted till the critical line coincides with the zero point of the micrometer scale. The prism is then cleaned and filled with the specimen under test. The position of the critical line on the scale now indicates the refractive index of the specimen.

On the eyepiece micrometer are engraved two horizontal scales. The upper one applies to specimens at 22° C., and the lower one to specimens at 45° C. The index of refraction of a liquid at 22° C. corresponding to any scale reading d on the upper scale is

$$n_{22} = 1.4675 \pm 0.00025\,d,$$

while that of a liquid at 45° C. corresponding to any scale reading d on the lower scale is

$$n_{45} = 1.4595 \pm 0.00025\,d,$$

where the plus sign is to be used when the deviation is to the right of the zero line of the scale, and the negative sign is to be used when the deviation is to the left.

49. The Dawes Refractometer. — This instrument of recent design consists of a Gauss eyepiece G, and a piece of plane parallel-sided glass in contact with a double convex lens. A drop of the liquid whose refractive index is required is placed between the glass plate and lens. The back surface of the lens is silvered.

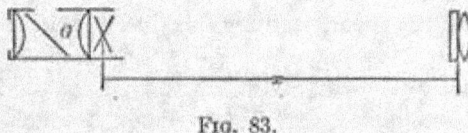

FIG. 83.

A Gauss eyepiece is one with a side opening and a piece of unsilvered glass inclined 45° to the axis. Light reflected by this mirror strongly illumines the cross hairs without interfering with the view through the eye-lens.

In the case of the present instrument light diverging from the cross hairs as a source traverses the plane glass, the specimen and the lens, and is reflected back by the silvered surface. The setting of the instrument for finding the refractive index of the specimen consists in adjusting the distance x between the cross hairs and the plane surface of the specimen till the image of the cross hairs, formed by the reflected light, coincides with the cross hairs themselves.

In order that the cross hairs and their image may coincide, the two waves traversing any selected point on the axis must have equal curvatures. We will now set up an expression for the curvature of the wave front moving to the right, and also one for the curvature of the wave front moving to the left, when each is at the left pole of the lens. Equating these two values we will have an expression involving the required refractive index, the distance x, and measurable constants of the apparatus.

Let t' and n' represent the thickness and refractive index, respectively, of the glass plate. At the first surface of the glass plate, the radius of curvature of the wave front proceeding from

the cross hairs is $x - t'$. At the second surface, it exceeds this
value by the distance the wave travels in glass during the time it
would travel the distance t' in air. Whence, at the second surface of the glass plate, the radius of curvature of the wave front is

$$R = x - t' + \frac{t'}{n'}.$$

And since the curvature of a spherical surface equals the reciprocal of the radius of curvature, the curvature of the wave front at
the second surface of the glass plate is

$$C \left[= \frac{1}{R} \right] = \frac{1}{x - t' + \frac{t'}{n'}}. \qquad (146)$$

In traversing the plane surface separating the glass plate and
the specimen there is no change of curvature of the wave front.

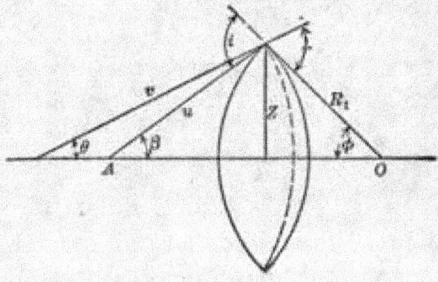

FIG. 84.

The change of curvature in passing through the spherical surface
between the specimen and lens will now be found. In Fig. 84
the heavy curve of radius R_1 represents the surface separating the
specimen of refractive index n from the lens of refractive index
n''; the light curve of radius n represents an incident wave front;
and the dashed curve of radius v represents the front of the corresponding refracted wave. From Snell's law of refraction

$$n \sin i = n'' \sin r.$$

We will use such a small lens aperture that in place of the sines of i and r we can put the angles themselves, expressed in radians. Then, the above becomes

$$ni = n''r.$$

From the geometry of the figure, $i = \beta + \phi$, and $r = \theta + \phi$. So that

$$n(\beta + \phi) = n''(\theta + \phi),$$
$$n''\theta - n\beta = \phi(n - n'').$$

Since the aperture of the lens is so small, the line z approximately equals the arc subtended by the angles θ, β, and ϕ. Whence, expressed in circular measure,

$$\theta = \frac{z}{v}; \quad \beta = \frac{z}{u}; \quad \phi = \frac{z}{R_1}.$$

Consequently, the above equation may be written

$$\frac{n''}{v} - \frac{n}{u} = \frac{n - n''}{R_1}. \tag{147}$$

The left member of this equation is the change in curvature impressed on the wave entering the surface separating the specimen from the lens.

Consequently, on entering the lens, the curvature of the wave front originating at the cross hairs, is (146) and (147),

$$\frac{1}{x - t' + \frac{t'}{n'}} + \frac{n - n''}{R_1}. \tag{148}$$

We will now find an expression for the curvature, at this same point, of the wave front which, reflected from the concave silvered surface, forms an image coinciding with the object. Call the radius of curvature of the silvered surface R_2, and the distance between the poles of the lens t''. As the image is to coincide with the object, the wave front incident on the mirror must have the same curvature as the mirror. On traveling back through a medium of refractive index n'' for a distance t'', the radius of curvature will be $\dfrac{R_2 - t''}{n''}$, and the curvature will be

$$\frac{n''}{R_2 - t''}. \tag{149}$$

Since in traversing this point in either direction, the curvature of the wave front is the same, we have, from (148) and (149),

$$\frac{1}{x - t' + \dfrac{t'}{n'}} + \frac{n - n''}{R_1} = \frac{n''}{R_2 - t''}. \tag{150}$$

This may be written

$$\frac{1}{x - t' + \dfrac{t'}{n'}} + \frac{n}{R_1} = \frac{n''}{R_1} + \frac{n''}{R_2 - t''},$$

or

$$\frac{1}{x - B} + \frac{n}{R_1} = A,$$

when A, B, and R_1 are constants for a particular instrument.

Whence, the required refractive index

$$n = \left(A - \frac{1}{x - B}\right) R_1 = C - \frac{R_1}{x - B}, \tag{151}$$

where C represents the constant quantity AR_1.

The three constants in the above equations are most easily determined from measurements of x for three substances of known refractive indices.

50. The Plane Diffraction Grating. — A transmission diffraction grating consists of a piece of clear glass on which are a large number of fine equally spaced parallel straight opaque lines a few wave-lengths apart. Such a grating may be made by plowing furrows through the surface of a thin piece of plate glass with a diamond point. Replicas of a diamond ruled grating can be cheaply made either by photography or by taking a cast of it in collodion.

A reflection diffraction grating consists of a highly polished metal surface in which have been ruled a large number of equally spaced straight grooves a few wave-lengths apart. Transmission gratings can be made by flowing a thin solution of collodion over a reflection grating, removing the dried collodion film and mounting it on a glass plate. Gratings usually have from ten to twenty thousand lines per inch.

Consider, first, the effect of a transmission grating on a plane wave incident *normally* upon it. In Fig. 85 are represented three transparent spaces of a transmission grating. Every point of each space is a center of disturbance from which light proceeds in every direction. At first suppose the incident light is monochromatic. From the center of each space represented in the

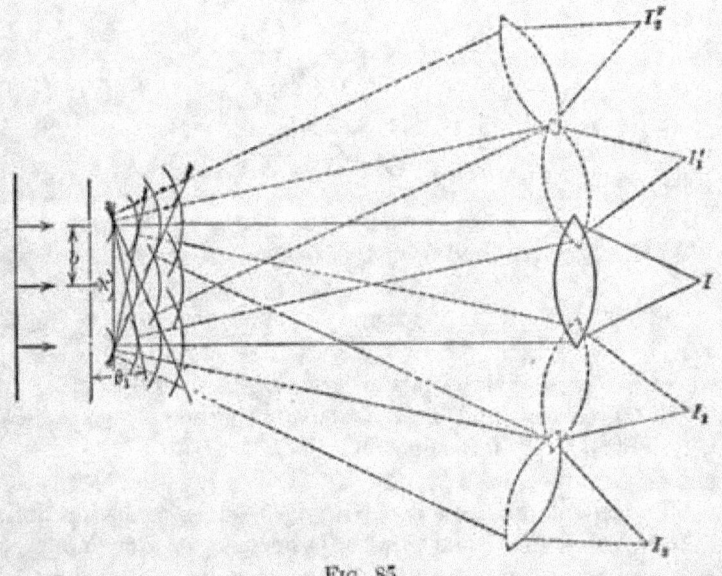

Fig. 85.

figure are drawn several spherical wavelets, one wave-length apart. Surfaces tangent at every point to these wavelets are wave fronts advancing from the grating. In the figure several such plane wave fronts are represented. Two rays from each wave front are indicated. If a positive lens be placed in the path of the light from the grating, each of these plane waves will be converged to a focus. That is, each of these plane waves will produce an image of the object from which comes the light incident on the grating. Besides the central image I, there is a series of images I_1, I_2, I_3, etc., and another series I_1', I_2', I_3', etc.

If instead of being monochromatic, the light source emits light of several wave-lengths, each of the images will be multiple, that is, will consist of an image of the object for each wave-length. At the central image I, the various colored images are superposed producing a single resultant image of the color of the object. At I_1, I_2, I_3, etc., and I_1', I_2' I_3', etc., the separate images of various colors are side by side, the spaces between the separate images depending upon the differences between the wave-lengths of the components of the incident light. The multiple images, other than the central one, are called spectra. The spectra at I_1 and I_1' are called the spectra of the "first order," those at I_2 and I_2' are called the spectra of the "second order," etc. Beyond the spectra of the second order formed by gratings of more than about fifteen thousand lines per inch, there is considerable overlapping of the spectra of various orders.

Light Incident Normally on a Transmission or on a Reflection Grating. — The effect of a reflection grating on which light is incident normally can be also obtained from a study of Fig. 85. In this case the grooves take the place of the opaque spaces of the transmission grating, reflecting spaces take the place of the transparent spaces, and the light is incident from the right normally.

If we represent the width of one grating element, that is, the width of one transparent space together with one opaque space (or, in the case of a reflection grating, one reflecting space together with one groove) by the symbol b, and the angle between the grating and a wave front which gives an image of the first order by the symbol θ_1, we have from Fig. 85,

$$\lambda = b \sin \theta_1,$$

where λ is the wave-length of the light in the particular image.

In general, if the angle between the grating and the wave front which gives an image of the nth order be represented by θ_n,

$$\lambda = \frac{b \sin \theta_n}{n}. \tag{152}$$

The diffraction grating furnishes the simplest method for the accurate determination of wave-lengths of light. The dispersion varies inversely with the grating space b; the resolving power varies directly with the total number of lines in the ruled space.

Plane gratings are usually mounted at the center of a divided circle as shown in Fig. 136. Light from the source after traversing the slit S and lens L_1 constitutes a parallel beam incident on the grating G. The slit S and lens L_1 constitutes a system called a collimator. After leaving the grating — either by transmission or reflection — the light is focalized by the lens L_2 in the plane of the cross hairs D. This image is magnified by the lens L_3. The lenses L_2 and L_3 with the cross hairs constitute a reading telescope.

A reflection grating can be set normal to the direction of the incident light as follows: First with the grating removed take the reading of the pointer attached to the telescope when the

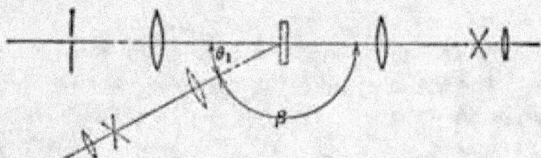

FIG. 86.

image of the slit is at the intersection of the cross hairs. Then, after placing the grating in place with the reflecting surface toward the collimator, take a reading of the pointer when the first order image to the right is on the intersection of the cross hairs. The angle β through which the telescope has been turned is the complement of the deviation θ_1 of the first order image on the right-hand side. In the same manner find the deviation θ_1' of the first order image on the left-hand side. When the plane of the grating is normal to the direction of the incident light, θ_1 will equal θ_1'.

Light Incident Obliquely on a Reflection Grating. — In Fig. 87, let AB and $A'B'$ be two positions of a small part of the grating capable of rotation about O. Let ϕ be the angle between the two positions of the grating, and θ the angle between the axes of the collimator and the telescope. Let XY be one grating space.

Draw OP and YZ perpendicular to AB and OQ perpendicular to $A'B'$. CY and $C'X$ are parallel rays from the collimator, and YT and XT' are parallel rays from the grating to the telescope.

Now it is a fact if light be reversed in direction it will retrace its original path. Thus in the present case, if the collimator and

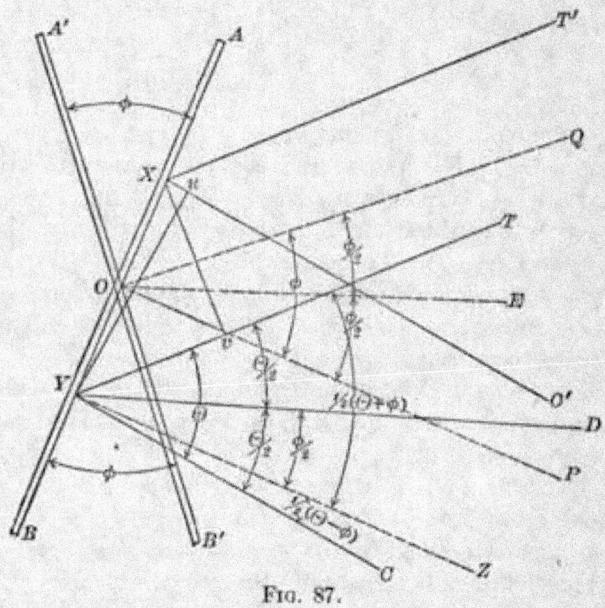

FIG. 87.

telescope be interchanged, light incident on the grating in the direction TY would enter the telescope in the direction YC. It follows that OE, the bisector of the angle QOP, is parallel to YD, the bisector of the angle TYC. And since YZ is parallel to OP, and $POQ = \phi$, it follows that $ZYD = \tfrac{1}{2}\phi$.

Draw Yu perpendicular to XC', and Xv perpendicular to YT. Yu is the wave front on incidence, and Xv is the wave front on reflection. This reflected wave front is in advance of the incident wave front by the distance $(Yv - Xu)$. The condition for a bright line in the spectrum is that this distance shall equal a whole number of wave-lengths. That is,

$$n\lambda = Yv - Xu,$$

where n is the order of the spectrum.

But, from the construction of the figure,

$$Yv = YX \sin YXv = YX \sin TYZ = YX \sin \tfrac{1}{2}(\theta + \phi)$$

and $Xu = YX \sin XYu = YX \sin CYZ = YX \sin \tfrac{1}{2}(\theta - \phi),$

denoting the distance YX between two consecutive rulings by b,

$$n\lambda\ [= Yv - Xu] = b \sin \tfrac{1}{2}(\theta + \phi) - b \sin \tfrac{1}{2}(\theta - \phi)$$

$$= 2b \cos \frac{\theta}{2} \sin \frac{\phi}{2},$$

or
$$\lambda = \frac{2b}{n} \cos \frac{\theta}{2} \sin \frac{\phi}{2}, \qquad (153)$$

where n is the order of the spectrum.

In applying this method, the collimator and telescope are clamped together at a known angle θ. The grating is turned till the spectral line whose wave-length is required is on the cross hairs of the telescope. The grating is then turned till the same line from the same order of spectrum is obtained from the other side of the normal to the grating. The angle between these two positions of the grating is the angle ϕ of (153).

51. The Concave Diffraction Grating. — By means of a grating ruled on a concave reflecting surface, a sharply defined spectrum can be formed without the aid of lenses. The method of determining wave-lengths by means of a concave grating will now be developed.

In Fig. 88, let G represent the grating surface with center of curvature at C and the rulings normal to the plane of the diagram. Suppose that X and Y are two adjacent rulings. (As a matter of fact, the distance between two adjacent rulings is only about 0.0002 cm.) Construct a circle through Y having the radius of the grating as a diameter. At some point S on this circle put the slit S parallel to the rulings. At some other point I on the same circle, light diffracted from X and Y will produce a bright image of the slit. For wave-length λ, this will occur when

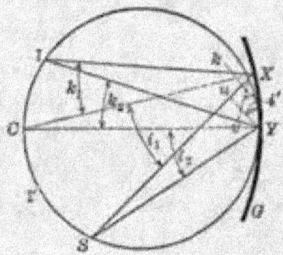

Fig. 88.

$$n\lambda = uX - vY.$$

Since X and Y are very close together, the angles of incidence i_1 and i_2 are very nearly equal, and the angles of diffraction k_1 and k_2

are very nearly equal. Consequently, we will represent each angle of incidence by the same symbol, i, and each angle of diffraction by the same symbol, k. From the figure, since uY is perpendicular to SX, and XY is perpendicular to CY, $uYX = i$. And since vx is perpendicular to IY, and XY to CY, $vXY = k$. Whence, $uX = XY \sin i$, and $vY = XY \sin k$. Substituting these values in the above equation,

$$n\lambda = b (\sin i - \sin k), \tag{154}$$

where b represents the distance XY between two consecutive rulings.

At the center of curvature C of the grating, light of wave-length λ_c will produce a bright image of the slit. For this position, $k = 0$, $\sin k = 0$, and (154) becomes

$$n\lambda_c = b \sin i. \tag{155}$$

At some point I', light of wave-length λ' will produce a bright image of the slit. For any image between C and S,

$$n\lambda' = b (\sin i + \sin k). \tag{156}$$

Consequently, the higher orders of spectra, and the spectral lines of greater wave-lengths of any one order, lie nearer the slit.

52. Spectra. — When a parallel beam of unhomogeneous light from a narrow slit falls upon a prism or grating the component colors of the beam are dispersed. If a converging lens be placed in the beam after leaving the prism or grating, light of each frequency will form a separate image of the slit. If the frequencies of the various components differ considerably one from another, the separate images of the slit will be distinctly separated. Such a group of images of the slit is called a "bright line spectrum." But if the differences in the frequencies from one component to the next are small, the various images of the slit will merge into one another thereby forming a "continuous spectrum." An incandescent gas free of solid particles gives a bright line spectrum. An incandescent solid or liquid gives a continuous spectrum.

If in a beam of light giving a continuous spectrum there be interposed a transparent colored substance, certain colors of the continuous spectrum will be blotted out. The resulting spectrum which is essentially a continuous spectrum interrupted by dark spaces is called an "absorption spectrum."

The positions of the lines of a bright line spectrum, and of the dark bands of an absorption spectrum, are characteristic of the substance producing the spectrum.

Wave-lengths of light are expressed in various units. The micron, represented by the symbol μ, is 0.001 mm. The Ångström * unit, represented by the symbol A is 10^{-10} meters. For example, the wave-length of the light of the line D_1 of the sodium spectrum is

$$0.00005896 \text{ cm.} = 0.5896\,\mu = 5896 \text{ Å}.$$

53. The Spectroscope and the Spectrometer. — A spectroscope is an instrument for producing spectra. A spectroscope provided with a scale by means of which wave-lengths can be determined is called a spectrometer. Spectroscopes and spectrom-

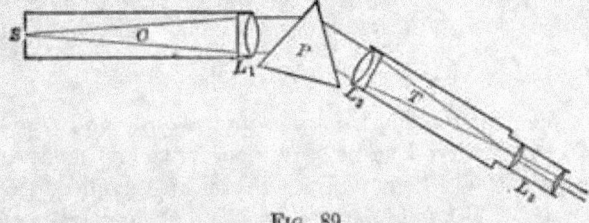

Fig. 89.

eters are made in an immense variety of forms. But in all types, light from the source under investigation, after traversing a narrow slit, and after being dispersed into its components by either a prism or a grating, forms an image of the slit for each wavelength.

A common type, Fig. 89, consists of a slit S at the principal focus of a lens L_1, one or more prisms (or a plane grating) at P, and an astronomical telescope T with a Ramsden eyepiece and cross hairs. The tube C with the slit at one end and the lens at the other is called the collimator. The purpose of a collimator is to render parallel the light diverging from the slit.

* Pronounced Ong'-stroem.

The "direct vision" spectroscope consists of a slit S, Fig. 90,
a lens L_1 to render parallel the light from the slit, a system of
crown glass and flint glass prisms by means of which the light is
dispersed without being deviated from its incident direction, a
lens L_2 which converges light of each frequency to a separate

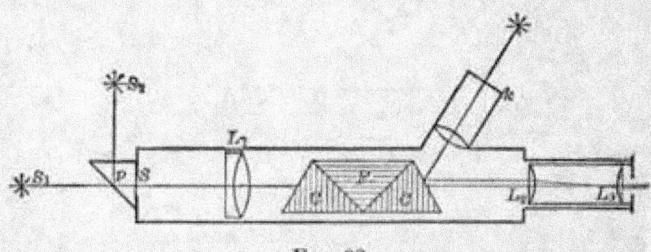

FIG. 90.

image of the slit, and a lens L_3 for magnifying this spectrum.
Light may enter the slit from a source in the axis of the instrument, or, if a totally reflecting prism P be employed, from a
source at one side of the instrument. This totally reflecting
prism usually covers but one-half the length of the slit so that
light from two sources can enter at the same time. By this
device, which can be applied to a spectroscope of any type, the
spectra of two substances are formed side by side. If the wave-
lengths of the lines of one spectrum are known, the wave-lengths
of the lines of the other spectrum can be estimated by a comparison of the distances between the lines. This instrument is
often provided with a scale k ruled on glass. When this scale is
properly illumined, an image of the scale is formed in the image
plane. The positions of the spectrum lines can then be referred
to this fixed scale.

In the "auto-collimating" spectroscope the functions of both
collimator and telescope are performed by the telescope. In the
focal plane of the telescope, Fig. 91, is a diaphragm filling one-
half of the cross section of the tube. In this diaphragm is the
slit s which by means of the total reflection prism p is illumined
by the source S. Light from the source, after traversing the slit,
diverges to the objective O from which it emerges in a parallel

beam and is incident on the dispersing prism P; is reflected from
the posterior silvered face; returns through the prism; under-
goes an additional dispersion on emergence; returns through the
objective and forms an image of the slit for light of each wave-
length in the half of the focal plane not covered by the diaphragm.

With all prism spectroscopes, the images of the slit are some-

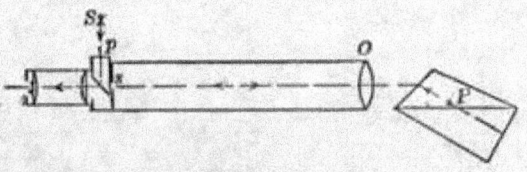

Fig. 91.

what curved, the rays from the ends of the slit being more deviated than those from the middle. If the deviation is great, as
when a train of prisms is used, the curvature is very marked. The
light from the ends of the slit suffers greater deviation than that
from the middle because it traverses the prism in directions inclined to a plane normal to the refracting edge of the prism.

For the study of the parts of a spectrum below the visible red
and beyond the violet, care must be taken that energy of these
particular frequencies is not absorbed by transmission through
the parts of the apparatus. The prism and lenses of instruments
designed for the study of the infra-red region can be made of rock
salt, sylvanite, or fluorite, while those of instruments designed for
the ultra-violet region quartz is employed. Or, the optical system may be entirely reflecting so that no energy is lost by absorption through transmitting media. In this case concave mirrors replace the lenses, and a diffraction grating takes the place
of the prism.

The reflection spectrometer can be reduced to its lowest terms
by ruling the grating on the surface of a concave mirror. With
the concave diffraction grating no lenses and no other mirrors are
required. The radius of curvature of concave gratings is usually
6, 10, or 21.5 feet. The Rowland mounting, which is the one
usually employed, is described in Exp. 45.

54. Qualitative Spectrum Analysis. — The grouping of lines in the emission spectrum of any element is characteristic of that element and is the same whether the element is alone or mixed with others. After a survey has been made of the spectra of all elements, the elements present in any given specimen can be determined by matching the lines appearing in the spectrum of the specimen with those of the elements. This fact is the basis of qualitative spectrum analysis. The sensitivity of the method is so great that 10^{-10} gm. of calcium or 10^{-11} gm. of strontium can be detected.

To vaporize a substance various methods are available. Some substances can be vaporized in the flame of a Bunsen burner. Others require an electric arc or a spark discharge. Some lines of a given substance are made visible by a source at one temperature, while other lines of the same substance are made visible only at a considerably higher temperature. In the Bunsen flame, strontium chloride gives lines due to strontium chloride molecules, bands due to the oxide, and one line due to the metal strontium. The arc and the spark spectra consist of about 40 lines, all due to the metal itself. Some elements give two or more entirely different spectra depending upon the means used in rendering them luminous. For example, the spectrum of nitrogen at low pressure produced by an ordinary spark discharge consists of bands, whereas when considerable electric capacity is introduced in parallel with the spark gap, an entirely different spectrum is produced consisting of sharp lines. Oxygen has three spectra. The introduction of capacity in parallel with the spark gap produces an oscillatory discharge which is of higher current value than the discharge without the capacity. (Reference — *Baly, Spectroscopy*, 1905, pp. 550, Longmans, Green & Co.)

55. Quantitative Spectrum Analysis. — When white light traverses a colored substance placed in front of the slit of a spectroscope an absorption spectrum is formed. In the case of solutions, the amount of weakening in any particular part of the spectrum depends upon both the thickness of the layer traversed by the light and the concentration of the solution. Whenever the absorption of light is a constant function of the concentration of the solution, the value of the concentration can be determined from the absorption. Absorption at any part of the spectrum is proportional to concentration so long as there is no chemical or physical change in the solution. It should be kept in mind, however, that some substances in aqueous solution become less hydrolyzed at greater concentrations, for example, the chlorides and bromides of copper, cobalt, iron, and nickel. Again, the absorption spectrum of a solution of a solute when not ionized may be different from that of the same solute when ionized. Changes in temperature also produce changes in the absorption produced by certain substances in aqueous solution, for example, the salts of iodine, cobalt, and chromium.

The absorption spectrum of a solution containing two or more solutes may have bands due to each solute separate from the bands due to spectra of the other solutes. In such a case the concentration of each solute can be determined as though the others were not present. However, if the bands instead of being separate are superposed, the quantitative analysis is usually impossible.

We will now derive one of the relations used in the determination of the concentration of solutions. If the intensity of the light incident upon a layer of unit thickness be I, then the intensity of the emergent light will be aI, where a is a constant having a value less than unity. On traversing a second layer of equal thickness, the intensity of the emergent light is $a(aI) = Ia^2$. And on traversing n similar layers, that is, a thickness n, the intensity I' of the emergent light is

$$I' = Ia^n,$$

or
$$a = \left(\frac{I'}{I}\right)^{\frac{1}{n}}. \tag{157}$$

The consideration will be simplified by expressing the quantity a in terms of a layer of such thickness N that $I' = 0.1\,I$. Then the above equation becomes

$$a = (0.1)^{\frac{1}{N}}.$$

The quantity $1/N$, that is, the reciprocal of the thickness of specimen required to reduce the intensity of the emergent light to one-tenth that of the incident light is called the *extinction coefficient*. Representing the extinction coefficient by the symbol E, the preceding equation becomes

$$a = (0.1)^E,$$

or
$$\log a = E \log 0.1 = -E,$$

but from (157)
$$= \frac{1}{n} \log \frac{I'}{I}.$$

Whence,
$$E = -\frac{1}{n} \log \frac{I'}{I}. \tag{158}$$

Since the absorption of a solution is proportional to the concentration and to the thickness of layer, it follows that if the thickness be constant the concentration is directly proportional to the extinction coefficient. Thus we may write

$$\frac{c}{c_1} = \frac{E}{E_1}. \tag{159}$$

Thus the concentration of a solution can be found from a determination of the ratio I'/I. The absorption of light of any particular frequency can be determined from a photometric comparison of two lights of the same frequency one of which has traversed the solution.

The spectrophotometric method for the determination of the concentration of solutions is susceptible of a high degree of precision. For example, the concentration of solutions of Cu ions, made by dissolving a known mass of pure metallic copper, of concentrations from 0.25 per cent to 2.50 per cent, can be readily determined with an error of less than 0.05 per cent. For very dilute solutions the method is much more precise than chemical analysis. For certain classes of substances the method offers considerable advantage as to speed. For example, after the preliminary curves have been made for solutions of known concentration, it is possible to estimate quantitatively dilute solutions of lead, calcium, ammonia, sulphates, and chlorides, in from 15 to 20 minutes, with an accuracy far beyond any chemical method.

The estimation of the concentration of solutions by means of the intensity of the absorption bands of their spectra is often called spectrocolorimetry. (Reference — *Krüss, Kolorimetrie und quantitative Spectralanalyse*, pp. 291, Leipzig, 1891.)

56. Colorless Solutions. — The concentration of a solution that does not absorb light cannot be directly determined by spectrophotometric means. But by adding a reagent that either frees a colored ion, produces a colored salt, or develops a turbidity depending upon the concentration of the solution, the value of the concentration can be determined. For example, potassium iodide is a colorless salt whereas the iodine ion is violet. This ion can be liberated by Cl, HNO_2, or $K_2Cr_2O_7$ in acid solution. The last reagent, however, would be inadmissible on account of the color added by the reagent itself. The following additional examples illustrating the method employed to produce either a color or a turbidity in a colorless solution are taken from a baccalaureate thesis presented to Purdue University by Mr. George Spitzer.

For the estimation of the concentration of a salicylic acid solution, ferric chloride was added, thereby forming a colored solution of ferric salicylate.

For the estimation of a dilute solution of ammonium hydrate add Nessler's solution. The solution becomes yellow due to the nitrogendimercurousiodide formed.

For the estimation of the concentration of a solution of ammonium nitrate, add sulphuric acid together with phenol and ammonia. The solution becomes yellow due to dissolved ammoniumnitrophenol.

For the quantitative estimation of phosphorus in fertilizers, first dissolve the phosphates by adding sulphuric acid. Neutralize by adding ammonia. Then form color by adding ammonium molybdate solution. In dilute solution the color is a light transparent yellow. In more concentrated solutions there is a yellow precipitate. This degree of concentration should be avoided.

In the determination of the concentration of magnesium chloride, add sodium acid phosphate. The turbidity produced by the formation of insoluble magnesium phosphate is a measure of the quantity of magnesium in the original solution.

For the estimation of calcium in a solution of calcium chloride, add ammonium oxalate. The turbidity produced by the formation of insoluble calcium oxalate is a measure of the quantity of calcium in the original solution.

Cane sugar can be determined as quickly and with as great precision by means of a spectrophotometer as by means of a polarimeter. The cane sugar is first turned into invert sugar by the addition of hydrochloric acid and heat. After making this slightly alkaline it is added to sufficient boiling alkaline cupric tartrate (Fehling's solution) to make a faint blue color. During this operation cuprous oxide is formed in proportion to the amount of cane sugar. The cuprous oxide is filtered, washed, dissolved in nitric acid, and diluted with water to a definite volume. After heating the solution to expel nitric oxide gas, the concentration of cuprous oxide in the solution is determined by means of the spectrophotometer. From empirical tables one can find the amount of cane sugar corresponding to this amount of cuprous oxide.

57. The Spectrophotometer. — An instrument by means of which the illumination at different parts of a spectrum can be measured is called a spectrophotometer. Spectrophotometers are used for the measurement of the amount of light of various frequencies emitted by a luminous source, and also the amount of light of the various frequencies absorbed by a given substance.

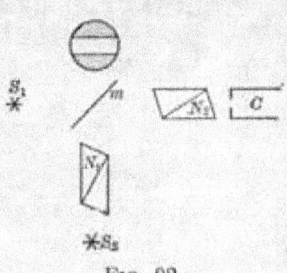

Fig. 92.

A spectrophotometer consists essentially of a spectroscope arranged for the passage of two beams of light, to which is added a device for comparing the brightness of the light of any given frequency in the two beams. The comparison of the brightness is usually made by means of a pair of polarizing prisms.

A convenient spectrophotometer can be made by placing in front of the slit of the spectroscope collimator, C, Fig. 92, two

Nicol prisms N_1 and N_2, and a plane glass mirror m, from which has been removed a horizontal stripe of silver. Light from the source S_1 will traverse the transparent stripe, the Nicol prism N_2 and the spectroscope; while light from the source S_2, after traversing the Nicol prism N_1, and being reflected from the silvered surface of m, will be transmitted by the Nicol prism N_2 and the spectroscope. In the image plane of the spectroscope will appear the spectrum of the source S_2 crossed horizontally by the spectrum of S_1. By means of a diaphragm with a vertical slit placed in the image plane, the attention can be fixed on two adjacent bright patches of the same frequency, one due to light from one source and one due to light from the other.

It will now be shown that the ratio of two illuminations produced by lights of the same frequency can be determined from the angle between the planes of polarization when the two parts of the field of view are equally bright. If the Nicol N_2, Fig. 92, were not present, all of the light in the focal plane of the telescope which originated at S_2 would vibrate in some plane op, Fig. 93, with an amplitude represented by the length ox. If the Nicol N_2 were in place but the Nicol N_1 were absent, all of the light in the focal plane which

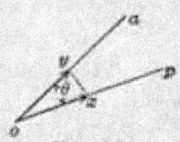

Fig. 93.

originated at S_2 would vibrate in some plane oa. If both prisms were in place, the light, which on leaving N_1 vibrated in the plane op and had an amplitude ox, would emerge from N_2 with the vibration in the plane oa and have an amplitude

$$oy = ox \cos \theta.$$

Thus, the ratio of the amplitudes of vibration of light that has traversed two polarizers to that which has traversed one equals the cosine of the angle between the planes of polarization of the two beams of light.

And since the energy of a vibration of constant frequency varies as the square of the amplitude, the ratio of the illuminations of the light which has traversed two Nicols to that which has traveled one is

$$\frac{I_2}{I_1}\left[= \frac{(oy)^2}{(ox)^2}\right] = \cos^2 \theta. \tag{160}$$

Consequently, by rotating either Nicol, thereby changing θ, we can vary by a known amount the illumination in the focal plane produced by the source S_2. Now the illumination in the focal plane due to light from S_1 is unaffected by rotation of either Nicol. Therefore, by rotating either Nicol till the two light patches in the focal plane are equally bright, we can find a numerical measure of the intensity, for the given frequency, of the source S_2 compared with that of S_1.

When the circular scale attached to the rotatable Nicol is so arranged that the zero reading indicates that the planes of polarization of the two Nicols are parallel, the above equation can be employed. Sometimes, however, the scale is arranged so that the zero reading indicates that the planes of polarization of the two Nicols are perpendicular to one another. In this case, calling the angular reading θ, we have

$$\frac{I_2}{I_1}[=\cos^2(90-\theta)]=\sin^2\theta. \tag{161}$$

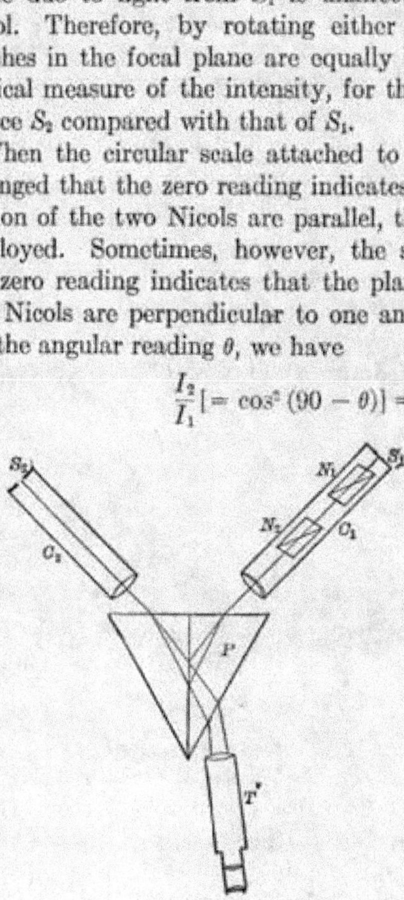

FIG. 94. 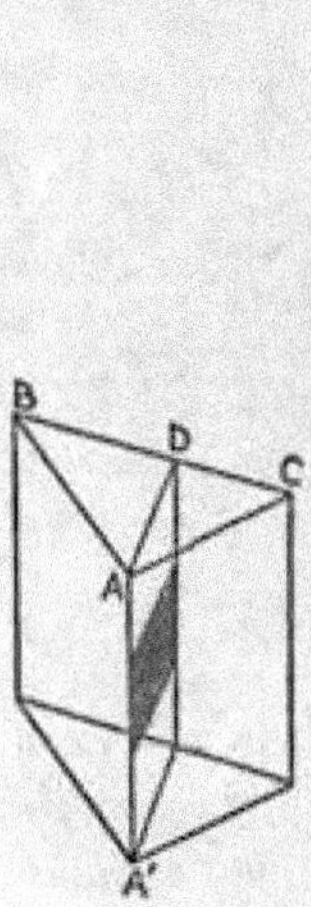   FIG. 95.

58. The Lemon-Brace Spectrophotometer. — This instrument* consists of two collimators C_1 and C_2, Fig. 94, a telescope T, and a Brace prism P. This prism is equiangular and is divided

* The Astrophysical Journal, Vol. 39, p. 204.

into two parts by a plane bisecting the edge AA', Fig. 95. The two inner faces are polished, and on one is deposited a horizontal stripe of silver. The two halves of the prism are then cemented together.

Light from the source S_1, Fig. 94, is reflected from the silver surface and forms a narrow horizontal spectrum of S_1 in the focal plane of the telescope. Light from the source S_2 traverses the prism above and below the silver stripe, and forms two horizontal spectra of S_2 in the focal plane of the telescope. The field of view thus consists of spectra of the two sources, side by side and in the same plane. The positions of the collimators can be so adjusted that in any vertical line in the field of view the light from the two sources will be of the same wave-length. The attention can be fixed on a limited portion of the two spectra of the same wave-lengths by means of an eyepiece diaphragm containing a vertical slit.

One collimator is provided with a pair of Nicol prisms N_1 and N_2. By rotating one Nicol, the parts of the two spectra uncovered by

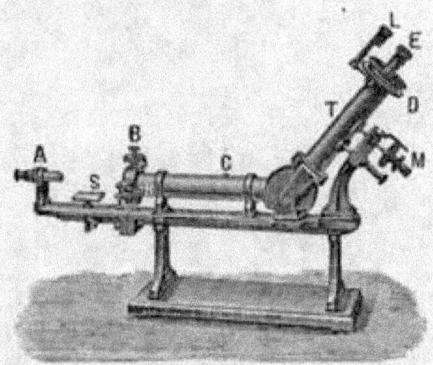

Fig. 96.

the diaphragm can be brought to the same brightness. After a photometric balance has been obtained, the ratio of the illuminations produced at the collimator slits by S_1 and S_2 can be obtained from (160) or (161). This is also the ratio of the luminous intensities of the two sources.

59. The Martens-Koenig Spectrophotometer. — This instrument,* Fig. 96, is especially suited to the spectrophotometric

* Annalen der Physik, Vol. 12, p. 984.

study of absorptive media. The optical system is shown in plan,
and in elevation in Fig. 97. It is essentially a two-lens prism
spectroscope with a photometric device depending upon polarization. The slit and the refracting edge of the dispersing prism
are horizontal. Cemented to each lens is an acute-angled prism
p and p' which prevents light reflected from the lens surfaces to

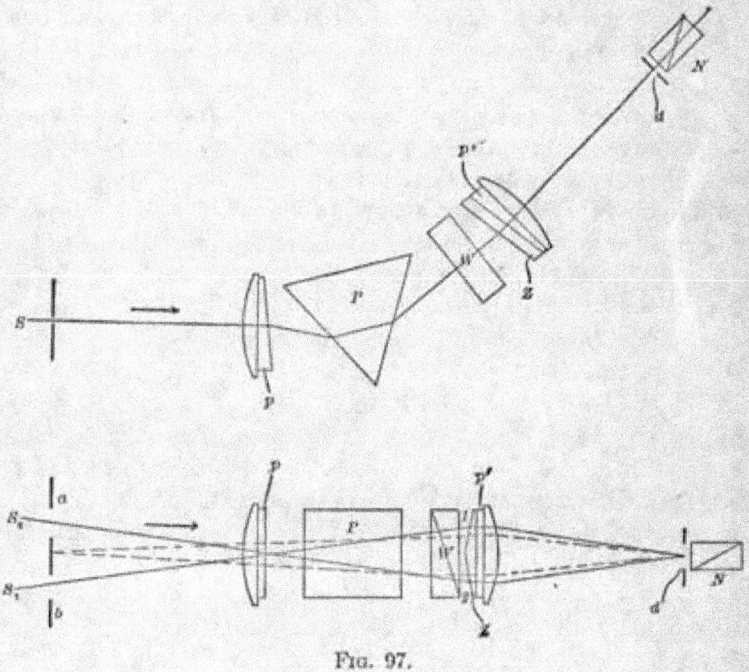

Fig. 97.

obscure the field of view. In the diagram, W is a polarizing device called a Wallaston prism, N is a Nicol analyzer, Z is a glass
biprism of very obtuse angle. The entrance slit is divided by a
tongue into two slits, a and b, through which pass two luminous
pencils S_1 and S_2.

The collimating lens renders each pencil parallel. The prism
P disperses each pencil into its component colors. The Wallaston prism divides each pencil into two plane polarized pencils,

— one with the plane of polarization vertical and the other horizontal. Each of these four pencils fills the two inclined faces of the biprism Z. On emergence from this prism there are eight dispersed plane polarized beams, four due to light from one slit and four due to light from the other. The lens cemented to p' focalizes each of these beams into a spectrum in the plane of the diaphragm d.

The changes produced in the two beams of light as they traverse the optical system may be illustrated by the diagram, Fig. 98. Throughout the diagram, the light that entered at the slit b, Fig. 97, is distinguished by the symbol "1" and that which entered at a by the symbol "2." A polarized beam with vibrations in the horizontal plane is distinguished by the subscript h, and that with vibrations in the vertical plane by the subscript v. If the Wallaston prism W and the biprism Z, Fig. 97, were not present, there would be formed in the plane of the diaphragm d two spectra of unpolarized light 1 and 2, Fig. 98. With the Wallaston prism in the path of the beams giving rise to these spectra, each spectrum is divided into two spectra of polarized light 1_v, 1_h, 2_v, and 2_h. All four beams giving rise to these spectra impinge on both faces of the biprism Z. The parts of the four beams that strike the upper face of the biprism will be refracted downward. On emergence these parts are distinguished in the diagram by "primes." The parts of the beams that strike the lower face will be bent upward. On emergence these parts are distinguished by "seconds." All light spectra thereby formed are focalized in the plane of the diaphragm D. The biprism is so tilted that the two rows of spectra instead of being superposed are side by side. In Fig. 98, the two rows of spectra $2_h'$, $2_v'$, $1_h'$, $1_v'$ and $2_h''$, $2_v''$, $1_h''$, $1_v''$ are shown in different planes. Actually they are side by side in one plane perpendicular to the page.

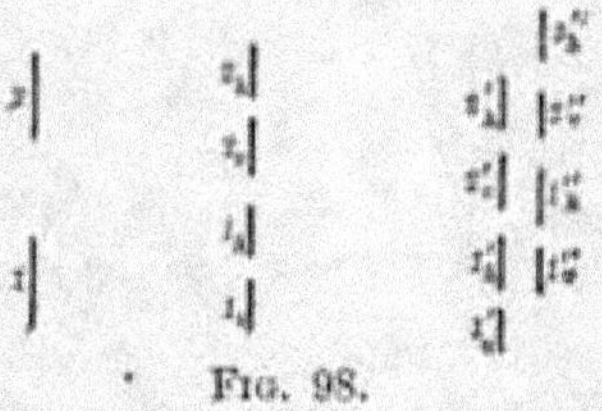

In the eyepiece diaphragm d there is a slit, parallel to the slits a and b, which cuts off all the light except a narrow strip from each of the adjacent spectra. One-half of the slit is illumined by

FIG. 98.

monochromatic plane polarized light from one source, while the other half is illumined by light of the same frequency polarized in the plane at right angles to the first from the other source. By interposing a Nicol prism N between the diaphragm and the eye, the two halves may be brought to equal brightness. The frequency of the light can be altered by rotating the observing tube about a horizontal axis through the dispersing prism.

For the spectrophotometric comparison of two light sources, the instrument is provided with two right-angled prisms x and y,

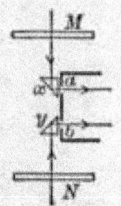

Fig. 99, by means of which light from two sources can be reflected into the slits a and b in the direction of the axis of the collimator. By means of ground glass diffusing screens M and N, the illumination of the slits is rendered uniform throughout their length.

Fig. 99.

The ratio between the luminous intensities of two light sources for any particular frequency will now be determined. Denote the luminous intensities for any particular frequency of the sources, sending light through the slits b and a by J_1 and J_2, respectively, and the illuminations at the diaphragm d by I_a and I_b, respectively. Then,

$$I_a = k_1 J_1, \tag{162}$$

and
$$I_b = k_2 J_2, \tag{163}$$

where k_1 and k_2 are constants depending upon the position of the luminous sources with respect to the diffusing screens, and the reflecting and absorbing powers for light of the particular frequency of the diffusing screens and optical system of the instrument.

The two patches of light at the diaphragm d due to the two sources are plane-polarized in planes at right angles to each other and are of the same frequency. On looking at the diaphragm through a Nicol prism one sees the circular field of view divided into two halves by a sharp line. In one position of the Nicol, one-half of the field will appear dark and the other bright, while if the Nicol be rotated 90° the half formerly dark will be bright and the half formerly bright will be dark. At an intermediate position the two halves of the field of view will be equally bright. In Fig. 100, let OM be the plane of polarization of light in the half of the field of view coming from the slit a, and let ON be the plane of polarization of the light from the slit b. Let OH and OC represent the amplitudes of vibration of the light constituting these two halves of the field of view. Let OQ be the plane of transmission of the Nicol prism when the two halves of the field of view are of equal illumination. This equality requires that the angle θ be such

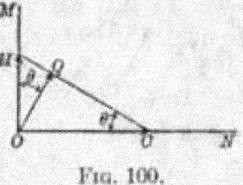

Fig. 100.

that the projection of OH on OQ equals the projection of OC on OQ. That is, OQ is perpendicular to HC. It follows that

$$OG = OH \cos \theta = OC \sin \theta.$$

That is, the amplitudes of vibration of the light in the field of view of the instrument coming from the slits a and b are in the ratio

$$\frac{OH}{OC} = \tan \theta.$$

And since intensity of illumination varies directly with the square of the amplitude of vibration,

$$\frac{I_a}{I_b} \left[= \frac{(OH)^2}{(OC)^2} \right] = \tan^2 \theta. \tag{164}$$

From (162), (163), and (164),

$$\frac{I_a}{I_b} = \frac{k_1 J_1}{k_2 J_2} = \tan^2 \theta. \tag{165}$$

That is, for a particular frequency, the ratio of the luminous intensities of the two light sources is

$$\frac{J_1}{J_2} = k_3 \tan^2 \theta. \tag{166}$$

60. Colorimetry. — If the extinction coefficient of a solution were the same for light of all frequencies, the relative concentration of two specimens of the same solution could be obtained from a

direct comparison of the intensity of color of the light transmitted
by them when illumined by white light of the same intensity. But
there are no substances that have a constant extinction coefficient
through a wide range of frequencies. That is, a comparison of
color intensities will not give accurate results when applied to
solutions which absorb light through a wide range of frequencies.
In case, however, the absorption be limited to a single narrow
region of the spectrum, the light emerging from two specimens of
the solution of different concentrations will show the same depth
of color when the thickness of the two specimens traversed by
the light is inversely proportional to their concentrations. This

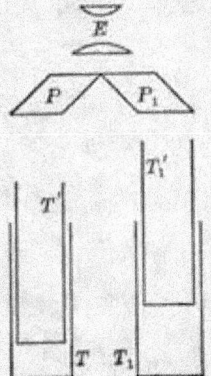

Fig. 101.

method of comparing the concentrations of
two specimens of the same solution is called
colorimetry.

In the Dubosc colorimeter, the two specimens are contained in the vertical tubes T and
T_1, Fig. 101, provided with plane glass bottoms.
The thickness of the layers traversed by light
is varied by moving up or down the inner tubes
T' and T_1' provided with plane glass bottoms.
By means of a mirror below the apparatus,
light is sent through the two specimens. After
traversing the specimens, the light illumines the
upper faces of the total reflecting prisms P and
P_1. On looking through the eyepiece, E, one
sees two patches of colored light separated by a sharp line. The
thickness of each layer is given by a scale attached to the containing
tube. When the thickness of the layers have been adjusted till the
two halves of the field of view are of the same depth of color, the
concentrations of the two specimens are inversely proportional to
their thicknesses.

With this instrument, when the concentrations are not too different, settings can be repeated with a maximum departure from
the mean of about 2 per cent. The definiteness of the balance
between the two halves of the field of view can be improved by
adding a difference of hue to the difference of brightness. For
example, suppose two concentrations of a bluish solution are being

compared. On interposing between the eye and the eye-lens a sheet of glass or gelatine tinted yellow, the two halves of the field of view will appear green when they are of equal brightness. But if one-half be brighter than the other, this will be yellowish-green, and the darker half of the field of view will be bluish-green.

The concentration of a solution is expressed in several ways. The per cent concentration by weight and the per cent concentration by volume are often used. The concentration is also expressed in terms of a molar solution and in terms of a normal solution. A molar solution is one containing one molecular weight of the substance, in grams, to one liter of solution. A normal solution of a reagent is one which contains in one liter that proportion of the molecular weight of the reagent, in grams, which corresponds to one gram of available hydrogen. For example, a molar solution of $CuSO_4 + 5\,H_2O$, in water, contains $63.57 + 32.07 + 64 + 5\,(2 + 16) = 249.64$ gm. of copper sulphate and enough water to make one liter of solution. A normal solution of the same substance contains $\frac{1}{2}$ (249.64) gm. of copper sulphate per liter of solution. Again, a normal solution of Na_2SO_4 is of one-half the concentration of the molar solution. A normal solution of $NaHSO_4$ is of the same concentration as the molar solution.

61. Rotation of the Plane of Polarization. — When plane polarized light is incident on certain crystals or on certain carbon compounds, the plane of polarization of the emergent light is not that of the incident light. Some substances rotate the plane of polarization about the axis of the beam in the clockwise direction as viewed by an observer looking toward the approaching light, while others rotate it counterclockwise. The former substances are called right-handed, dextrogyric, or positive, and the latter substances are called left-handed, levogyric, or negative.

It is found that the amount of rotation produced by a given substance is directly proportional to the thickness, and depends upon the temperatures of the specimen and upon the wave-length of the incident light. In the case of solutions of active substances in inactive solvents, the rotation is proportional to the concentration. When light traverses more than one substance, the rotation equals the algebraic sum of the rotations due to the separate substances.

The rotation, in degrees, of a decimeter length of solution, divided by the concentration in grams per 100 cc., is called the *specific rotation* of the solute at the given temperature, for light of the given wave-length. Thus, if 10 grams of a certain sugar be made into an aqueous solution of 100 cc. and the rotation produced by a layer 20 cm. thick be 13.3°, the specific rotation of the sugar at the concentration and temperature of the measurement and with the particular wave-length of light employed is

$$[a] = \frac{(13.3)\ 100}{(2.0)\ 10} = 66°.5.$$

Usually, specific rotations are given at 20° C. and for sodium light.

The specific rotations of various sugars at 20° C., and at $t°$ C., at concentration c, with sodium light, are as follows:

Sucrose,　　　$[a]_{20} = 66.51 + 0.0045\ c.$　　　　　　　$[c = 5\ \text{to}\ 65.]$
　　　　　　　$[a]_t = [a]_{20} - 0.0144\ (t - 20).$
Dextrose,　　　$[a]_{20} = 52.5 + 0.0188\ c.$　　[No change with temp.]
Levulose,　　　$[a]_t = -88.13 - 0.2583\ c + 0.6714\ (t - 20).$
Invert sugar,　$[a]_t = -19.8 - 0.036\ c + 0.304\ (t - 20).$
Lactose,　　　$[a]_{20} = 52.53.$　　　　[No change with concentration.]
　　　　　　　$[a]_t = [a]_{20} - 0.07\ (t - 20).$
Maltose,　　　$[a]_t = 140.37 - 0.0184\ c + 0.095\ t.$

From the definition of specific rotation it follows that the rotation due to a solution of an active substance in an inactive solvent is

$$\theta = \frac{[a]\ lm}{v} = \frac{[a]\ lc}{100}, \tag{167}$$

where a is the specific rotation, l is the length of solution traversed by the light, m is the mass of substance, v is the volume of solution, and c is the concentration usually expressed in grams of substance per 100 cc. of solution.

The angle of rotation of the plane of polarization of a substance is different for light of different wave-lengths. This variation of the rotatory power with the wave-length of the light is called *rotatory dispersion.*

The phenomenon of the rotation of the plane of polarization is extensively applied in the identification of rotary active substances, and in the determination of the concentration of solutions of such substances. (Reference — Landolt, *The Optical Rotating Power of Organic Substances and its Pretical Applicaations*, pp. 751, The Chemical Publishing Co., Easton, Pa.)

62. Half-shade Polarimeters. — An instrument for measuring the angle of rotation of the plane of polarization is called a polarimeter or polaristrobometer. The obvious method of determining the angle of rotation produced by any specimen would be to place the specimen between two Nicol prisms set for extinction and then rotate one prism till the field of view again becomes dark. The trouble with this simple method is that the eye is not very sensitive to small changes of brightness, and the mind cannot accurately compare the brightness of two things unless seen simultaneously and in juxtaposition. In the most sensitive method of measuring the angle of rotation there is added to the two Nicol prisms a device so arranged that the field of view is divided into either two or three parts which are of equal brightness for a certain position of the analyzer and which are of much different brightness when the analyzer is but slightly rotated from this position. On account of the low brightness of the field at this sensitive point, the method is called the *half-shade* method.

In the half-shade method, the plane polarized light traversing the specimen consists of two beams with their planes of vibration inclined to one another at a small angle. Without an analyzing Nicol, both halves of the field of view will be of the same brightness. But with an analyzing Nicol, the two halves will be unequally bright except when the principal plane of the analyzer bisects the angle between the planes of vibration of the two beams incident on the analyzer.

63. Laurent's Half-shade Polarimeter. — In this instrument the device used to produce the separation of plane polarized light into two portions consists of a plate of quartz YXY' and a plate of glass $YX'Y'$, Fig. 102, joined together along one edge. The plate of quartz is cut with the optic axis parallel to the joint YY', and is of such a thickness that during the passage of light through it, light in the extraordinary ray is retarded more than light in the ordinary ray by an amount equal to one-half wave-length of the monochromatic light used. The glass plate is of such a thickness that the light which traverses it is reduced in brightness, through absorption and reflection, by the same amount as the light that traverses the quartz plate.

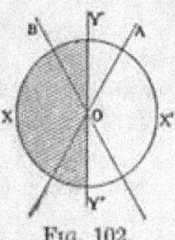

FIG. 102.

Suppose the monochromatic plane polarized light incident on the compound quartz-glass plate vibrates parallel to some line OB. The part of the light incident on the glass plate emerges vibrating in the same plane OB, but the portion incident on the quartz emerges as plane polarized light vibrating in some other plane OA. It can be shown, though the proof will not be here given, that the planes of vibration of the emergent light, OB and OA, are equally inclined to the joint YY'. Consequently, when a Nicol prism is placed in front of the quartz-glass plate with the principal plane parallel to the joint YY', the field of view is uniformly bright. With the Nicol turned out of this position, even very slightly, the two halves of the field of view are of very unequal brightness. This quartz-glass plate is called Laurent's half-shade analyzer. The angle BOA between the planes of vibration of the two plane polarized beams emerging from the half-shade analyzer is called the half-shade angle.

As usually employed, the half-shade analyzer is placed between two Nicol prisms with the joint between the quartz and glass plates slightly inclined to the principal plane of the polarizing Nicol. The polarizing prism is illumined with monochromatic light and the analyzing prism is turned till the field of view is uniform. The specimen under investigation is then placed between the half-shade analyzer and the analyzing prism. If a rotation of the plane of polarization has been produced, one-half the field of view is now dark and the other is bright. The angle the analyzing Nicol must be turned to bring the two halves to equal brightness is the amount of rotation produced by the specimen.

64. The Lippich Half-shade Polarimeter. — The only essential difference between the Lippich and the Laurent polarimeters is in the half-shade analyzer. The Lippich half-shade analyzer consists of either one or two supplementary Nicol prisms placed between the polarizer and analyzer of the instrument. The optical system of the Lippich half-shade polarimeter with triple field is represented in Fig. 103. After being rendered nearly monochromatic by traversing the light filter F, the light traverses the illuminating lens O and polarizer P. On emerging from the polarizer, one-third of the beam enters the prism L_1, one-third enters the

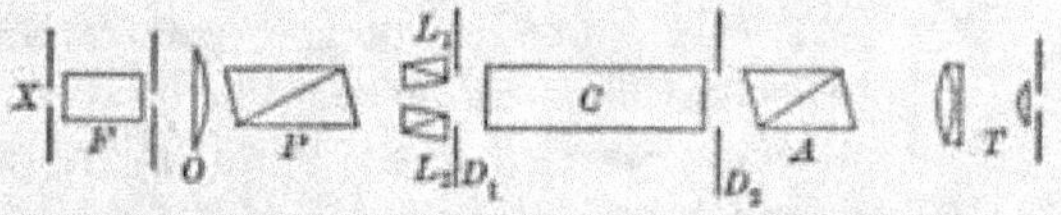

Fig. 103.

prism L_2, and the remainder proceeds directly to the diaphragm D_1. Thus, three beams traverse the specimen C, the analyzer A, and the telescope T. The field of view consists of a circle divided into three stripes of equal area and sharply marked from one another by two sharp lines.

The principal sections of the prisms L_1 and L_2, constituting the Lippich half-shade analyzer must be nearly coplanar. The angle between the principal plane of the Lippich analyzer and that of the polarizer is the half-shade angle. To produce variations in sensitivity, this half-shade angle is changed by rotating the polarizer. When the light source and analyzer diaphragm D_2 are at conjugate foci of the illuminating lens O, the polarizer diaphragm D_1 is uniformly illumined. The eyepiece is focalized on the polarizer diaphragm.

The light in the middle stripe of the field of view is polarized at right angles to the principal section of the polarizer P. The light in the outer stripes is polarized at right angles to the principal sections of the prisms L_1 and L_2. Thus, the plane of polarization of the light in the middle stripe and that in the outer stripes are inclined to one another at the half-shade angle. If there were no

light lost by reflection or absorption in traversing the Lippich
analyzer, then when the principal section of the analyzer A bisects
the half-shade angle, the entire field of view would be uniform, and
when the analyzer A is turned however slightly from this position
the middle stripe would have a very different brightness than the
outer stripes. Since, however, there is a loss of light in traversing

the Lippich analyzer, it follows that
in the zero position, the principal
section of the analyzer A does not
bisect the half-shade angle.

On account of its great sensitivity
and ease of manipulation, the Lippich
half-shade triple-field polarimeter is
in extensive use. One pattern, designed by Landolt for the examina-
tion of specimens contained in tubes
or vessels of any shape, is illustrated
in Fig. 104. A specimen tube C is

Fig. 104.

shown in position for observation. Another, enclosed in a constant
temperature device J, is shown in the foreground. By means of
the lever h, the polarizing Nicol can be rotated, thereby altering
the half-shade angle, and consequently the sensitivity of setting.
The analyzer A can be rotated by means of the lever H, or,
after tightening the clamping screw k, by means of the screw b.
The reading on the scale S is made by aid of the verniers vv and
reading lenses LL.

65. The Quartz-wedge Compensation. — As light of different
frequencies is rotated to a different extent by optically active
substances, the polarimeters above described require the use of
monochromatic light. This requirement may be obviated by the
addition of a simple device. This device consists of a set of dextro-
and levorotatory quartz wedges of acute angles placed between
the crossed Nicols of the polarimeter and so arranged that different
thicknesses of either dextro- or levorotatory quartz may be interposed in the path of light. The double quartz-wedge system used
by Schmidt and Haensch consists of two wedges, A and B, Fig. 105,
each capable of being moved up or down between two fixed

wedges C and D. One movable and one fixed wedge are levorotatory, and the other two are dextrorotatory, as indicated.

With the wedges in the positions shown in the diagram, the system gives zero rotation. By moving either A or B the system can be made either levo- or dextrorotatory. In this manner, the rotation produced by a specimen between the crossed Nicols of a polarimeter can be neutralized. The amount that A or B must be displaced from the zero position to produce compensation is a measure of the rotation of the plane of polarization produced by the specimen.

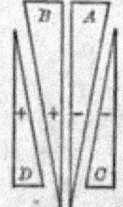

Fig. 105.

A quartz-wedge compensator cannot be used in testing a specimen having a rotation dispersion much different from that of quartz. As quartz has nearly the same rotation dispersion as has cane sugar, some form of quartz-wedge compensator is usually employed on saccharimeters, that is, on polarimeters designed especially for the testing of sugars. The effects due to the slight differences in the rotation dispersions of sugar and quartz are overcome by the use of a potassium-bichromate light filter.

66. Sugar Scales. — A saccharimeter is a polarimeter with a scale graduated to indicate directly the per cent of pure sugar contained in the dissolved sample. When the temperature, length of specimen tube, mass of sugar sample, and volume of aqueous solution are constant, the rotation varies only with the purity of the sample. If, in addition, the sugar sample contains no active substance other than sucrose, and if the weight of the sample be that which a specimen of pure sucrose would need to have to give a reading of 100 on the saccharimeter scale, then the reading for the given sample will give directly the per cent sucrose in it.

In saccharimeters of French design, the 100-degree point is the scale reading due to the rotation of the plane of polarization of sodium light produced by a plate of dextrorotatory quartz one millimeter thick cut perpendicular to the optic axis. Experiment shows that about 16.29 gm. of sucrose dissolved in water to 100 cc. at 20° C., in a tube 200 mm. long, gives the same rotation as 1 mm. of quartz. This number, 16.29 gm., was adopted in 1896 by the

International Congress of Applied Chemistry as the "normal weight" for the French Sugar Scale.

In most saccharimeters made outside of France the Ventzke Sugar Scale is employed. The International Sugar Commission in 1900 adopted the following definition of the 100-degree point of this scale. "The 100-degree point of the saccharimeter scale is obtained by polarizing a solution containing 26.000 gm. of pure sucrose (weighed in air with brass weights) in 100 true cc. at 20° C., in a 200-mm. tube in a saccharimeter whose quartz-wedge compensation must also have a temperature of 20° C." The white light must be filtered through a 3-cm. layer of 3 per cent potassium-bichromate solution.

The relations between one degree angular measure, one degree French sugar scale, and one degree Ventzke sugar scale, for sodium light, are as follows:

$$1° \text{ Angular} = 4.61553° \text{ French} = 2.88542° \text{ Ventzke.}$$
$$1° \text{ French} = 0.21666° \text{ Angular} = 0.62516° \text{ Ventzke.}$$
$$1° \text{ Ventzke} = 0.34657° \text{ Angular} = 1.59960° \text{ French.}$$

67. Saccharimetry. — When but one active substance of known specific rotation is present in a specimen, the percentage of this substance in the specimen can be determined from a single polarimetric observation. This is called the simple polarization method.

When a specimen contains more than one sort of sugar, it is possible to determine the percentage of one of the sugars provided that the rotatory power of this sugar can be altered a known amount. This method involves two polarimetric observations and is called the method of double polarization. There are three important means by which the rotatory power of certain sugars can be altered a definite amount. They are, (a), change of one sort of sugar into another by the action of acids or alkalis; (b), change of one sort of sugar into a different material by fermentation; (c), change of the rotatory power of one component by change of temperature.

(a) When hydrochloric acid is added to an aqueous solution of sucrose (cane sugar), each molecule of sucrose is changed into two molecules — one of d-glucose (dextrose) and one of fructose (levulose). The resulting mixture rotates the plane of polarization in the direction opposite that which does sucrose, and is called "invert sugar." The process of forming invert sugar is called "inversion." In the same manner, lactose is changed into equal parts of galactose and d-glucose, while maltose is turned into d-glucose, with an accompanying change of rotatory power.

Again, when weak solutions of glucose, fructose, invert sugar, lactose, or maltose are heated for a sufficient length of time in presence of sodium hydroxide, they become inactive due to the formation of mixture of dextro- and levorotating sugars. Sucrose is unchanged by this treatment.

Of these actions, the most important in practical saccharimetry is the inversion of sucrose for the determination of the percentage of sucrose in a sample containing noninvertable sugars. Suppose that the rotation produced by the solution before the addition of the acid is θ, that by the noninvertable sugars is β, and that the rotation produced after inversion is θ'. Represent the specific rotations of sucrose and invert sugar by $[a_1]$ and $[a_2]$, respectively; the concentrations of the sucrose and invert sugar in the solution by c_1 and c_2, respectively; and the length, in decimeters, of the column of solution before inversion by l. Then, from (167), before inversion,

$$\theta = 0.01 \, [a_1] \, c_1 l + \beta. \tag{168}$$

In inverting sucrose, it is customary to add to the solution 10 per cent of acid, by volume. Thus, the invert sugar occupies 1.1 the volume occupied by the sucrose. In an experiment, the required correction is commonly made by the use, after inversion, of a tube having the length 1.1 l. Thus, after inversion,

$$\theta' = 1.1 \times 0.01 \, [a_2] \, c_2 l + \beta.$$

Thus the change of the rotation produced by the inversion of the sucrose is

$$\theta - \theta' = ([a_1] \, c_1 l - 1.1 \, [a_2] \, c_2 l) \, 0.01.$$

Now, the molecular weight of invert sugar, ($C_6H_{12}O_6 + C_6H_{12}O_6$), is 360, while that of sucrose, ($C_{12}H_{22}O_{11}$), is 342. Thus, the ratio of the concentration of invert sugar to the sucrose from which it was derived is

$$\frac{c_2}{c_1} = \frac{360}{342} = 1.053.$$

Substituting this value of c_2 in the preceding equation,

$$\theta - \theta' = c_1 l \left([a_1] - 1.053 \times 1.1 [a_2]\right) 0.01.$$

Whence, the concentration of sucrose in the solution

$$c_1 = \frac{\theta - \theta'}{l \left([a_1] - 1.158 [a_2]\right) 0.01}. \tag{169}$$

After the concentration c_1 has been determined, the concentration of any known noninvertable component can be found. Thus, representing by $[a_3]$ and c_3, respectively, the specific rotation and concentration of this component, we have from (167)

$$\beta = 0.01 [a_3] c_3 l,$$

and from (168)

$$\beta = \theta - 0.01 [a_1] c_1 l.$$

Whence,

$$c_3 = \frac{\theta - 0.01 [a_1] c_1 l}{0.01 [a_3] l}. \tag{170}$$

(b) Some yeasts ferment completely certain sugars and do not affect others. For example, the yeast *Saccharomyces cerevisioe* ferments completely d-glucose, d-fructose, d-mannose, d-galactose, sucrose, and maltose, but does not act upon l-xylose, l-arabinose, rhamnose, sorbose, and lactose: *Saccharomyces apiculatus* ferments completely d-glucose, d-mannose, and d-fructose, but does not act upon galactose, sucrose, maltose, and lactose. Thus, if a specimen contains one fermentable sugar in presence of other unfermentable sugars, the percentage of the fermentable component can be determined by observing the rotation before and after this component has been fermented.

(c) Changes of temperature affect greatly the value of the specific rotation of fructose (levulose), arabinose, and galactose; affects to a less degree that of maltose and lactose; and does not affect that of d-glucose (dextrose). Due to this effect, solutions of pure invert sugar at 90° C. are optically inactive. This fact is the basis of methods for determining the percentage of d-glucose mixed with sucrose in products that do not contain other sugars sensitive to changes of temperature. (Reference — *Browne, Handbook of Sugar Analysis*, pp. 787, 1912, New York, John Wiley and Sons, Inc.)

68. Clarification of Sugar Solutions. — Under even the best of conditions the field of view is so faint when a polarimetric setting is made, that care must be taken to have as much light as possible traverse the solution. For the clarification of sugar solutions there are three substances in common use. Alumina cream is prepared by adding alum solution to an excess of hot washing soda solution and collecting the precipitate. After washing this precipitate with boiling water it is mixed with water to the consistency of thin cream.

Basic acetate of lead solution is made by grinding in a mortar two parts of recently ignited litharge with one part of acetate of lead and enough water to form a paste. Boil in about half the mass of water, filter, and place in a well-stoppered bottle.

Sodium sulphite solution consists of 100 gm. of sodium sulphite in one liter of water.

To clarify a colorless cloudy solution containing about 20 gm. sugar per 50 cc. add about 3 cc. alumina cream and a drop of basic lead acetate. If the sugar solution is yellow add alumina cream as above and more basic lead acetate solution. In case the sugar solution is brown add 2 cc. sodium sulphite solution, and then add slowly lead acetate solution, constantly shaking till no more precipitate is formed.

After waiting till the precipitation is complete, dilute the solution to 100 cc. and filter. The sugar solution is now ready to fill the 2-decimeter specimen tube.

Exp. 26. Determination of the Illuminating Power of a Gas with a Bar Photometer by the Equality of Brightness Method

THEORY OF THE EXPERIMENT. — Read Arts. 26–29 and 31 (*d*). The illuminating power of a gas is the luminous intensity of a flame burning at a specified rate. Luminous intensities are expressed

in terms of the luminous intensity of the Hefner lamp or in terms
of the international candle. The international candle is $\frac{1}{9}$ of a
hefner. It is practically realized by a spermaceti candle $\frac{7}{8}$ inch
in diameter burning at the rate of 120 grains per hour. An ordinary paraffin candle burning at the rate of 8 grains per hour has
a luminous intensity of very nearly one international candle.

It is common American practice to require illuminating gas to
give a luminous intensity of not less than 16 candle power when
burning in a special Argand burner at the rate of 5 cubic feet per
hour under a pressure of 30 inches of mercury and at a temperature
of 60° F. Since a burner operates best for a given rate of flow of
gas at a definite pressure, it is necessary that all measurements be
made under conditions differing but slightly from standard conditions. A standard Argand burner is designed to be operated
under the above standard conditions. But if the gas be dry, and
the rate of burning and the pressure do not differ much from standard conditions, the volume that the gas actually consumed would
have had if the pressure and temperature had been standard can
be obtained by means of the fundamental law of perfect gases, as
follows: Denoting the observed temperature, pressure, and volume
by t, p, and v, respectively, we may write

$$pv = Rm \, (t + 460), \qquad (171)$$

where t is expressed according to the Fahrenheit scale.

Representing by v', the volume which the same gas would have
had under standard conditions, we may write

$$30 \, v' = Rm \, (60 + 460). \qquad (172)$$

Dividing each member of (172) by the corresponding member
of (171),

$$v' = \frac{520 \, pv}{30 \, (t + 460)}. \qquad (173)$$

If, however, a wet meter be used, the gas will be saturated with
water vapor and the above equation will be inapplicable. In this
case, find the value of v' by means of Table 12. An inspection
of this table shows that, *to change the volume of illuminating gas,
saturated with water vapor, and at any temperature and pressure, to*

the volume it would have had at 60° F., and 30 inches of mercury, subtract one per cent for each 4° F. above 60° F., and add one per cent for each 0.3 inch above 30 inches of mercury pressure.

MANIPULATION. — The apparatus used in this experiment includes a graduated bar, Fig. 106, at one end of which is a standard Argand burner connected to the gas supply through a pressure regulator, manometer, and meter. At the other end of the bar are two comparison candles. Sliding on the bar is a box containing

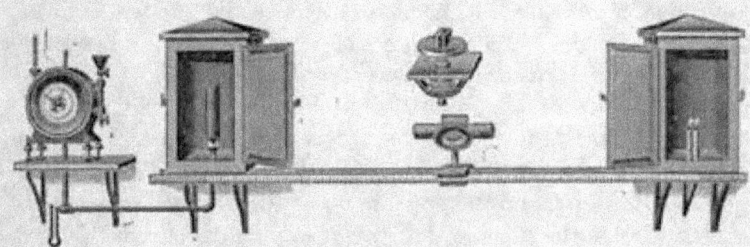

FIG. 106.

the photometer screen. In this experiment, the Joly cube will be found eminently satisfactory. This is shown in detail above the sliding box.

If the two blocks forming the Joly cube are not of equal thickness and scattering power, the position of equal brightness is determined by taking the mean of two scale readings when the blocks appear equally bright — first, when the blocks are facing in one direction, and then when the cube is rotated 180° about an axis through the plane of the tinfoil screen.

If one light source be much more intense than the other, the point of setting will be much nearer one source than the other. This diminishes the accuracy of the setting and also the degree of approximation with which (120) will apply. In the present experiment, two candles are to be used, close together and equally distant from the photometer screen.

See that the two sources of light are at the points corresponding to the ends of the scale of the photometer bar. As the bar usually does not extend as far as the end of the effective length of the bar,

this test is usually made by the aid of a pair of plumb lines at each
end of the scale, the plane of each pair being normal to the bar.
Examine the scale of the meter until you are certain you can read
it correctly. See that the water in the meter is at the height it had
when the meter was calibrated. With the stopcock at the burner
closed, turn on the gas at the meter, and see if the hand of the
meter moves. This will show if there be a leak between the meter
and burner.

Adjust the gas supply till, on lighting the lamp, there is a con-
sumption of very nearly five cubic feet of gas per hour. Then
adjust the draft of the lamp till the flame just ceases to tail over
the chimney. Light the candles and when they are burning freely,
extinguish them. Weigh them, together with the supports, to
within 10 milligrams. Replace the candles and relight them.
Record the hour, minute, and second, the meter reading, the barometric height, the water height in the manometer attached to the
pressure regulator, and the temperature as given by the thermometer in the gas meter.

Move the photometer screen back and forth till the two halves
of the block are equally bright. Record the scale reading. Rotate the photometer screen 180°, make a new setting, and record
the new scale reading. In the same manner, make at least five
pairs of readings. The reading of the scale is facilitated by reflecting on to it, by means of a mirror or white card, light from the
gas flame. During these measurements there must be no light in
the room except that due to the two sources being compared.

Extinguish the candles and the gas flame, recording the time in
hours, minutes, and seconds. Record the meter reading and re-
weigh the candles.

Note the readings of the barometer and the manometer, expressing both in the same units. The gas pressure equals the sum of
the barometric pressure and the pressure given by the manometer.

If a dry meter be used, compute, by means of (173), the volume
which the gas consumed in the given time would have occupied if
the pressure had been 30 inches of mercury and the temperature
had been 60° F. If, however, a wet meter be used, compute the

volume under standard conditions by means of the rule printed in italics at bottom of p. 174.

From the mass of paraffin burned, and the time occupied in burning, compute the candle power of the pair of candles. With this value, together with the readings on the photometer bar, compute, by means of (120), the actual candle power of the gas flame.

With this value, together with the previously computed value of the volume reduced to a pressure of 30 inches of mercury and a temperature of 60° F., compute the candle power which the same gas would have when burning at the rate of 5 cu. ft. per hour under the standard pressure and temperature. This is the illuminating power required.

Exp. 27. Determination of the Illuminating Power of a Gas with a London Gas Referees' Photometer

THEORY OF THE EXPERIMENT. — Read Arts. 26–29 and 31(e) and the preceding experiment. The gas referees of London, who supervise the daily tests of the gas supplied by all the gas companies of London, have developed an apparatus with which the photometric tests can be made quickly, and with sufficient precision, by untrained observers. To distinguish from the "bar photometer," Fig. 106, in which the photometer screen slides on a bar or track, this apparatus, Fig. 107, is often called the "table photometer." In this apparatus, the photometer screen is the photoped P, situated at a fixed distance of one meter from the Harcourt pentane 10-candle-power standard lamp PL. The gas under test is burned in a standard Sugg 24-hole Argand burner A. The distance between this burner and the photoped can be changed by means of the rod B. This rod is graduated so as to give directly the candle power of the gas flame when the two halves of the photoped are equally bright. The pressure regulator R and the gas meter M are of the wet type.

MANIPULATION. — Fill the tank or saturator of the pentane lamp two-thirds full with pentane. This must be done in a room in which there is no flame of any sort. While the lamp is in use, the depth of pentane in the saturator should never be less than

three millimeters. Adjust the lamp till the following conditions
are fulfilled. (a) The axis of the photometer passes through the
air tube of the lamp. (b) The center of the burner is over the
reference line of the photometer marked on the table. (c) The
distance from the burner to the chimney is 47 mm. (d) The mica
window is so placed that no light from it shall fall on any part of

Fig. 107.

the apparatus. (e) The bubbles in the spirit levels are in the
center of the vials. The valves on the saturator and on the de-
livery tube may now be opened, and if the gas will not ignite, the
draft must be forced by blowing into the funnel of the saturator.
Adjust the flame to a height of 1 cm. above the cross bars. Let
the flame burn steadily for 30 minutes before taking photometric
observations.

Meantime see that the water in the meter is at the height it had
when the meter was calibrated. Test for any leak between the
meter and the burner by closing the stopcock at the gas burner,
opening the supply stopcock, and observing whether the pointer
of the meter moves.

Regulate the flow of gas to 5 cu. ft. per hour. If the meter be
one such that one revolution of the pointer corresponds to one-

twelfth of a cubic foot, we will have a flow of 5 cu. ft. per hour when the pointer makes one revolution in one minute. In adjusting the flow of gas, take the time of three revolutions of the pointer by means of a stop watch. This time should not differ by more than three seconds from three minutes. Adjust the draft of the lamp till the flame is just on the point of smoking. The gas should burn steadily at least 15 minutes before taking photometric observations. The chimney must be clean.

In making a photometric setting, slide the rod carrying the Argand burner until the two halves of the field of the photoped are of equal brightness. Note the temperature and pressure of the gas. Correct the volume of gas to 60° F. and 30 in. of mercury as described in the previous experiment. Calling the corrected volume v', and assuming that illumination is proportional to v', the true candle power is obtained by multiplying the photometric scale reading by $5/v'$.

The normal density of pentane is 0.630 g. per cc. An increase in density of 0.01 g. per cc. increases the candle power of the pentane lamp about one per cent. Consequently, the saturator should be emptied at the close of the day's work. The stock of pentane should be stored in a metal tank with a tightly fitting screw plug.

Exp. 28. Calibration of a Carbon Incandescent Lamp to be used as a Working Standard

THEORY OF THE EXPERIMENT. — Read Arts. 26–29 and 31(f).

MANIPULATION. — In this experiment, the lamp being calibrated and a standard Hefner lamp are mounted on the ends of a horizontal divided bar on which slides a Lummer-Brodhun photometer. The brass base of the lamp being calibrated should be marked on the side toward the photometer so that in future use the same side of the lamp may be turned toward the photometer.

Adjust the height of the flame of the Hefner lamp until the image of the tip viewed by means of the eyepiece A, Fig. 57, coincides with the horizontal mark on the ground glass in front of the lens. After adjusting the flame, do not take photometric

observations for at least 15 minutes. Meantime, connect the
incandescent lamp in series with a storage battery and control
rheostat. Connect a suitable voltmeter across the lamp terminals.
Adjust the height of the Hefner lamp flame, the photometer
screen, and the incandescent lamp till their centers are at the
same height above the bar.

At an observed voltage, obtain the photometric balance. This
is best accomplished by moving the photometer back and forth
through the balance point by smaller and smaller displacements.
Any error due to lack of symmetry of the photometer is eliminated
by averaging the result obtained when the photometer is in one
position and that obtained when the photometer is rotated 180°
about a horizontal axis perpendicular to the bar.

Take a series of voltmeter and corresponding photometer readings extending from 30 per cent below to 20 per cent above normal voltage. Tabulate hefners, candle power, and volts. Plot a
curve coördinating candle power and volts.

In this experiment, there is a wide variation in the intensity
of one of the lamps, and consequently in the illumination of the
photometer screen at the moment a reading is made. This allows
some readings to be made with a greater precision than others.
A constant screen illumination at the time of a reading can be
obtained by clamping the carriage carrying the photometer and
one of the light sources to the ends of a rod of the proper length
to give the desired illumination. This coupled system is then
moved as a unit toward or away from the other light source till a
balance is obtained.

Exp. 29. Determination of the Mean Horizontal Candle Power and Mean Spherical Candle Power of an Incandescent Lamp

THEORY OF THE EXPERIMENT. — Read Arts. 26–29 and 31(f).
Except from a point or other symmetrical source, the distribution of light is not the same in all directions. The mean of the
luminous intensities in a horizontal equatorial plane, expressed
in candle power, is called the *mean horizontal candle power* of the

source. This may be obtained by averaging the candle power readings made at 10° or 15° intervals about the horizontal equatorial plane of the lamp. If the horizontal light distribution is not too ununiform, the mean horizontal candle power can be also obtained from a single reading made on a lamp rotating about its vertical axis at a speed of from 200 to 300 revolutions per minute.

If a sphere be imagined to be described about a source as center, and the surface of the sphere be divided into a large number of elements of equal area, and the candle power of the source be measured along the radius of the sphere through the center of each of these elements, the mean of the results so obtained is called the *mean spherical candle power* of the source. If the source be a lamp which can be rotated, — an incandescent lamp, for example, — the mean spherical candle power can be obtained from a much smaller number of observations. The theory of the method will now be considered.

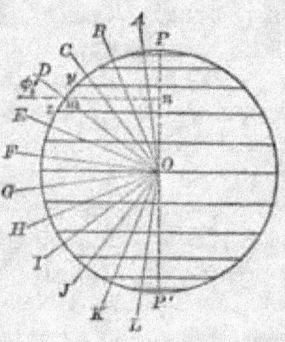

The luminous intensity of a source is, by definition (Art. 26), the luminous flux emitted by the source per unit solid angle. When the flux is ununiformly distributed, it will be convenient to take the sum of the amounts of flux traversing definite parts of an imaginary sphere having the source as the center. In Fig.

FIG. 108.

108, let the surface of such a sphere be divided into n narrow zones of equal angular width by planes normal to the axis of the lamp. We will now find an expression for the flux emerging from any zone along the radius of the sphere, in terms of the area of the zone and the mean luminous intensity of the source measured along a radius of the sphere passing through a midpoint of the zone. The total flux from all the zones, divided by 4π, *i.e.*, the number of steradians in a sphere, will give the mean spherical intensity of the source.

To fix the ideas, consider the zone of which m is a mid-point.

Represent by ϕ the angle between the radius OD passing through m and the base yz of the spherical segment of which the zone is the curved face. Representing by w_ϕ the solid angle subtended at the source by the zone, and by I_ϕ the mean luminous intensity of the source measured along radii passing through the zone, we have the value of the flux emitted by the zone

$$F_\phi = I_\phi w_\phi.$$

Representing the area of the zone by A_ϕ, and the radius of the sphere by r,

$$w_\phi = \frac{A_\phi}{r^2}.$$

The area of the zone equals the product of the mean circumference and the linear width measured along the arc. The mean radius of the zone is $mn = om \cos \phi$. The width is $2\pi r/2 n$. That is,

$$A_\phi = \frac{\pi r}{n} 2\pi r \cos \phi.$$

Whence, the flux emitted radially through the zone is

$$F_\phi [= I_\phi w_\phi] = \frac{2\pi^2 \cos \phi I_\phi}{n}.$$

And the flux from all the zones

$$F = \sum \frac{2\pi^2 \cos \phi I_\phi}{n}.$$

Therefore, the mean spherical candle power is, (117),

$$I_s \left[= \frac{F}{4\pi} \right] = \sum \left(\frac{\pi \cos \phi}{2 n} \right) I_\phi. \tag{174}$$

By directing a photometer toward the source along a line passing through a mid-point of a zone, the luminous intensity in that direction can be measured. If the source be spinning about the axis PP' with a speed of from 200 to 300 revolutions per minute, the photometer will indicate the mean luminous intensity I_ϕ throughout the particular zone. By taking such an observation along the line passing through a mid-point of each of the n zones, we can compute I, by means of (174).

Instead of maintaining the axis of rotation of the lamp fixed in space and directing the photometer along the lines $AO, BO,$ $CO,$ etc., it will be experimentally much easier to maintain the axis of the photometer horizontal, and change the axis of rotation, of the lamp so that $AO, BO, CO,$ etc., are successively in the axis of the photometer. The angle ϕ is then equal to the angular displacement of the axis of rotation PP' from the vertical, Fig. 109. A convenient rotator by means of which

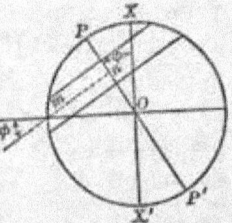

FIG. 109.

an incandescent lamp can be rotated about an axis inclined at any desired angle to the vertical is represented in Fig. 110. The spinning of the lamp is effected by means of a flexible cable extending from the lamp through a bent tube to a grooved pulley belted to a motor. The inclination of the axis of spin to the vertical is indicated by a divided circle shown in the figure.

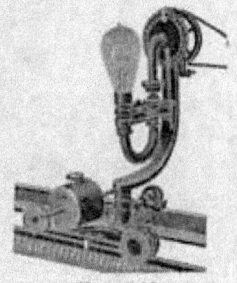

FIG. 110.

As an example of the use of (174), suppose that we take $n = 12$, that is, the angle at the center of the lamp subtended by each zone is $15°$. Then the angle between the axis of rotation and the line connecting the source and the mid-point of the first zone is $7° 30'$. Suppose that when the axis of rotation makes an angle of $7° 30'$ with the vertical, the photometric reading is 40 candle power. Then the first term of the series indicated in (174) is

$$\left(\frac{\pi \cos 7° 30'}{2 \times 12}\right) 40.$$

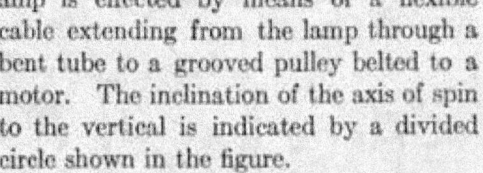

To facilitate the computation for a 12-zone sphere, the value of the quantity within the parenthesis of (174) corresponding to each of the 12 inclinations of the lamp is given below.

ϕ	$\dfrac{x\cos\phi}{2(12)}$	ϕ	$\dfrac{x\cos\phi}{2(12)}$
7° 30′	0.130	97° 30′	0.017
22° 30′	0.121	112° 30′	0.050
37° 30′	0.104	127° 30′	0.080
52° 30′	0.080	142° 30′	0.104
67° 30′	0.050	157° 30′	0.121
82° 30′	0.017	172° 30′	0.130

MANIPULATION. — In this experiment will be used a 2-meter bar with a Lummer-Brodhun photometer. The working standard will be a calibrated incandescent lamp maintained at constant voltage with the aid of a rheostat and voltmeter. Mount in the rotator the lamp under test and connect to a storage battery, rheostat, and voltmeter. The centers of the two lamps and the photometer screen are to be in line parallel to the photometer bar. In this experiment, the same side of the standard lamp should be directed toward the photometer that was so directed when the lamp was calibrated.

Adjust the potential differences at the terminals of the lamps till the two halves of the photometer screen are so nearly of the same color that there will be no difficulty in matching brightness. Thereafter, these potential differences must be constant. With the lamp under test rotating at the rate of from 200 to 300 revolutions per minute, make a photometric setting when the axis of rotation is vertical, and also when it is inclined to the vertical at angles of $7\frac{1}{2}°$, $22\frac{1}{2}°$, $37\frac{1}{2}°$, $52\frac{1}{2}°$, $67\frac{1}{2}°$, $82\frac{1}{2}°$, $97\frac{1}{2}°$, $112\frac{1}{2}°$, $127\frac{1}{2}°$ $142\frac{1}{2}°$, $157\frac{1}{2}°$, and $172\frac{1}{2}°$. Knowing the candle power of the standard lamp at the voltage used in this experiment, the candle power of the lamp under test in the above 13 directions can be computed.

Plot a polar curve showing the distribution of the mean luminous intensity, in a vertical plane, of the lamp under test. To construct this curve, draw from a center 24 radial lines 15° apart. Along these lines, lay off distances from the center proportional to the observed candle powers in the various directions. Join these end points by a smooth curve. The curve should be placed on the paper so as to show the distribution of luminous intensity of the lamp when the base of the lamp is above and the tip below.

Compute the mean horizontal candle power and the mean spherical candle power.

Exp. 30. Determination of the Candle Power of an Incandescent Lamp at Various Voltages by the Flicker Method

THEORY OF THE EXPERIMENT. — Read Arts. 26–28 and 30. After the temperature of an incandescent filament has reached a certain value, the luminous intensity increases rapidly with increase of temperature. That is, at higher temperatures the light emitted per unit of electric power is greater than it is at lower temperatures. For this reason, principally, metallic filament lamps that can be operated at high temperatures are now generally employed in preference to carbon filament lamps.

The object of this experiment is to measure the luminous intensity of a metallic filament lamp at a series of voltages extending from below the normal to above the normal voltage. The comparison lamp is to be another incandescent lamp maintained at constant voltage and which has been calibrated in terms of a Hefner lamp or a standard candle. As the colors of the two lamps will often be considerably unlike, the equality of brightness method of comparison will be unsuitable. For the present experiment will be employed a flicker photometer of the Bechstein type, Figs. 59 and 60.

MANIPULATION. — As in the equality of brightness method (Exp. 26), the sources to be compared are at the ends of a graduated track on which moves a carriage supporting the photometer. Adjust the resistance in circuit with the comparison lamp till the potential difference at the terminals of the lamp has the value it had when the lamp was calibrated. If the lamp under test be designed for 110 volts, adjust the resistance in circuit till the voltage is at first 95 volts.

Before starting the motor attached to the flicker photometer, move the photometer to the position in which the two parts of the field of view appear to be most nearly equally bright. This would be the setting if the equality of brightness method were to be employed. Now start the motor in slow rotation. The two parts of the field of the instrument flicker strongly. Gradually increase the speed of rotation till the flicker disappears. On moving the photometer back and forth on the track one will find that there is a short space in which no flicker occurs, and that when the photometer is outside this space there is flickering. By gradually diminishing the speed of the motor, the space of no flicker is shortened. At the proper speed the space of no flicker is reduced to a very short distance. After adjusting to this critical speed note the scale readings of the ends of the space of zero flicker. The mean of these readings is to be used in finding the values of r_1 and r_2 to be substituted in (120).

In the same manner take readings when the potential difference at the terminals of the lamp under test is 100, 105, 110, 115, 120, and 125 volts. With candle powers as abscissæ and potential differences as ordinates, plot a curve coördinating these quantities for the lamp under test.

Exp. 31. Determination of the Principal Focal Lengths, the Chromatic Aberration and the Longitudinal Spherical Aberration of a Converging Lens

THEORY OF THE EXPERIMENT. — Read Arts. 32–35. Longitudinal spherical aberration is measured by

$$\frac{f - f'}{f\theta^2},\tag{175}$$

where f and f' represent, respectively, the principal focal lengths for direct axial and for marginal axial pencils of white light, and θ represents the angle expressed in radians subtended at the image by the distance from the center of the lens to the center of the marginal pencils.

Chromatic aberration is measured by

$$\frac{f_r - f_b}{f},\tag{176}$$

where f_r, f_b, and f represent the principal focal lengths of the lens for direct axial pencils of red, blue, and white light, respectively.

For the determination of these four principal focal lengths, the corresponding principal foci and equivalent points must be located, and the distances between them measured. They can be located directly by means of an axial beam of parallel light. The easiest means for obtaining fairly monochromatic light is by interposing a piece of colored glass in a beam of intense white light. A sunbeam is parallel and intense but is not always available. The method to be employed in this experiment for producing an intense axial beam of parallel light will be made clear by reference to Fig. 111. The small hole O in a metal plate C is

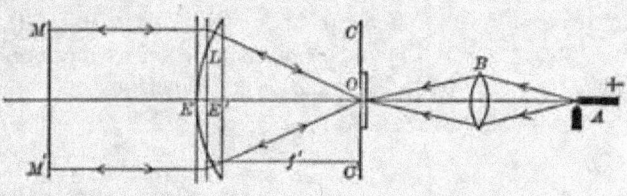

Fig. 111.

strongly illumined by the image of the crater of an arc lamp A. If the hole be covered by a piece of glass, either clear or colored, ground on the side adjacent to the metal plate, light will leave the hole as from a small bright object. After traversing the lens L under test, the light is reflected by the mirror MM'.

In the diagram, the equivalent points of the lens under test are indicated by E, E'. If the distance OE' equals the principal focal length of the lens, the emergent rays will be parallel to the principal axis of the lens. If the mirror MM' be normal to the principal axis, the reflected light will retrace its path, and the image of O will be formed at O. For the purpose of examining this image, it may be displaced to one side by slightly tilting the mirror.

MANIPULATION. — In the apparatus used in this experiment, Fig. 112, the lens holder H is mounted on a horizontal circular plate capable of rotation about a vertical axis. The lens holder H can be moved along a diameter of the plate by means of a rack and pinion T; and the plate with its attachments can be moved along a divided track as shown in the engraving.

By adjusting the lens B, focalize an image of the crater of the arc lamp A on the ground face of the colorless glass covering the aperture O. This gives us a brilliant small spot of light which

FIG. 112.

is to be used as the "object" in the subsequent measurements. Place close to the lens L under test a diaphragm D, containing a small central aperture. Adjust the inclination of the mirror M and the position of the lens carriage till there is formed on the white painted surface of C an image of the object O. Now place a piece of red glass over the aperture O and, by means of the rack and pinion T, adjust the position of the lens relative to the axis of rotation of the supporting circular plate till a slight rotation of the latter will cause no displacement of the image. During this adjustment the carriage must be moved so as to keep the image at its maximum sharpness. When this adjustment is complete, the distance from the axis of rotation of the circular plate to the screen C is the principal focal length f_r of the lens L for a direct axial pencil of red light.

Substitute for the diaphragm with the central aperture one that uncovers a concentric zone of the lens, and find the principal focal length of this zone of the lens. Repeat for the marginal zone.   During these two latter adjustments it is unnecessary to alter the position of the lens relative to the axis of rotation.

Measure the mean diameter of each of the zonal apertures. If r represents the mean radius of the zonal aperture, the angle at the image subtended by the mean radius of a zonal aperture

$$\theta = \tan^{-1}\left(\frac{r}{f_r}\right) \text{degrees} = \frac{1}{57.3}\tan^{-1}\left(\frac{r}{f_r}\right) \text{radians}.$$

These values of θ and the values of the focal lengths for central and marginal pencils substituted in (175) give the magnitude of the longitudinal spherical aberration of the lens when red light traverses it in the direction used in the experiment.   For a simple lens bounded by surfaces of unequal curvature, the magnitude of the longitudinal spherical aberration will be different when the direction of light through the lens is reversed.

Replace the diaphragm with the central aperture, substitute a piece of green glass for the red glass, and find the principal focal length f_g for the three apertures. Substitute a piece of blue glass for the previous piece, and find the principal focal length f_b for the three apertures. By means of (176) compute the magnitude of the chromatic aberration of the given lens.

Plot a curve coördinating longitudinal spherical aberration and mean radius of the zonal aperture.

Note. — If precise measurements are not required, but only the illustration of the accurate methods for the determination of the properties and constants of lenses and lens systems, much simpler apparatus can be used than that described in Exps. 31, 32, 33, and 35. For this purpose, the Farwell-Stifler Optical Bench, sold by the Standard Scientific Company of New York, will be found useful.

Exp. 32. Determination of the Principal Focal Length of a Diverging Lens

THEORY OF THE EXPERIMENT. — Read Art. 32. By definition, the principal focal length of a negative lens is the distance between the point from which an incident direct axial pencil of parallel light appears to diverge and the principal point of emergence of the lens. In Fig. 113, E_1' and E_1 are the equivalent points of the negative lens L_1 under test. If a direct axial pencil of parallel

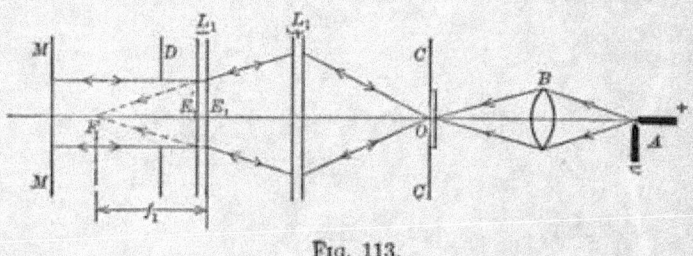

FIG. 113.

light be incident on the lens from the left, and if after emergence the pencil diverges from a point F_1, the principal focal length of the lens is $f_1 = F_1E_1$.

In the method to be employed in this experiment, the incident axial pencil of parallel light traversing the lens from left to right is obtained by the use of a small bright object O, a supplementary positive lens L_2, and a plane mirror M. By arranging the apparatus so that an image of the object O shall be formed on the screen C, which image shall remain stationary when the lens under test is rotated, the nodal point E_1 will be located. Since the lens is bounded on both sides by the same medium, this nodal point is the principal point of emergence for light traveling through the lens from left to right. If the lens L_1 under test be removed, an image of the object O will be formed at F_1 which can be located by means of a white screen. The distance F_1E_1 is the principal focal length required.

MANIPULATION. — Use is made in this experiment of the same Lens Testing Bench employed in the previous experiment. Place a piece of clear ground glass in the holder of the screen C with the ground side toward the screen. Adjust the lens B, Fig. 114, till the aperture O is strongly illumined by the image of the positive crater of the arc lamp A. Place the supplementary converging lens L_2 on the bench, and by means of a white card locate roughly the image of the object. Place between this image and the supplementary lens, the lens under test mounted in the rotatable carrier H. Put the mirror M in place, and slide L_1 and L_2 back and forth till an image of O is formed on the white surface of the screen C. Now, by adjustment of the rack and pinion T, and by motion of the carrier, find the position of

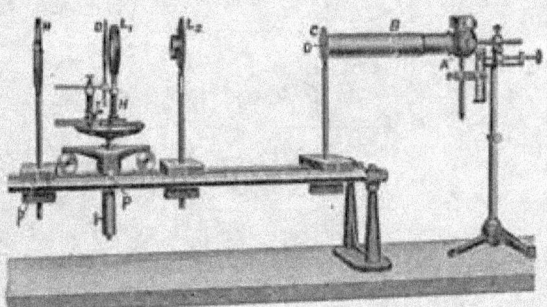

FIG. 114.

the lens under test such that if the lens be slightly rotated the image will not move. Note the scale reading of the pointer P. This indicates the position of the principal plane of emergence of the lens under test.

It will be noted that besides the image due to light that has been reflected from the mirror, there appears on the screen one or more other images due to light that has been reflected from the lens surfaces. The image due to light that has been reflected from the mirror may be identified by noting the image which moves when the mirror is slightly rotated.

Remove the diverging lens under test, turn the white back of the mirror toward the light source, and locate the image F_1, Fig. 113. Note the scale reading of the index p, Fig. 114. This indicates the position of the principal focus of the diverging lens under test. The difference between the readings of P and p is the principal focal length of the diverging lens under test.

Exp. 33. Location of the Equivalent Points and the Determination of the Principal Focal Length of a Lens System

THEORY OF THE EXPERIMENT. — Read Art. 32. The equivalent points of a lens system are located and the focal length determined exactly.as in the case of a single lens. One must first find whether the system is positive or negative. If positive, the method is as used in Exp. 31; if negative, a supplementary positive lens must be used as in Exp. 32.

In the first study of lens systems it will be convenient to use systems consisting of fairly large lenses arranged as in some of the standard eyepieces. The Ramsden eyepiece usually employed in instruments provided with cross hairs consists of two plano-convex lenses of equal focal length, with the convex surfaces facing one another and separated by a distance equal to two-thirds the principal focal length of one of them.

When cross hairs are not to be used, a system can be designed that will give less aberration than the Ramsden. The simplest Huyghens eyepiece consists of two plano-convex lenses, with the convex surfaces toward the light source, separated by a distance equal to the mean of their principal focal lengths. The principal focal length of the eye-lens is either one-third or one-half that of the field lens.

MANIPULATION. — By means of suitable clamps and rod, set up, as in Fig. 115, two lenses arranged as in a Ramsden eyepiece. Stop the field lens by a diaphragm provided with a central aperture. By examining the emergent beam of light with the aid of a white card one will find that the system is positive for light traversing it in either direction. Proceeding as in Exp. 30, locate an equivalent point and measure the principal focal length. Rotate the system 180° about a vertical axis, locate the other equivalent point, and measure the principal focal length.

Now set up, as in Fig. 116, two lenses arranged as in a Huyghens eyepiece. The lens of shorter focal length is the eye-lens and should be placed toward the screen *C*. The other lens is the field lens and should be placed toward the mirror. Stop the field lens by a

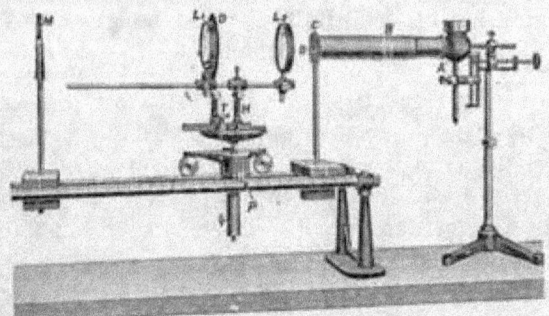

FIG. 115.

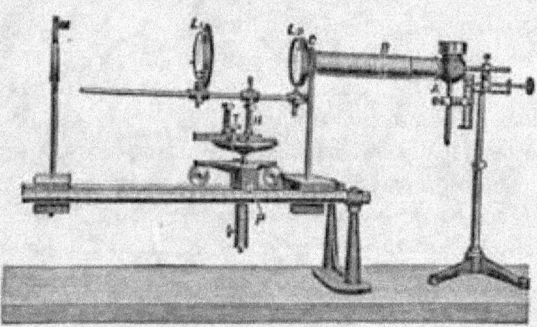

FIG. 116.

diaphragm provided with a central aperture. By examining the beam of emergent light with a white card we will find that the system is positive for light traversing it from the field lens to the eye-lens, Fig. 116, and negative for light traversing it in the reverse direction. When positive, locate the equivalent points and find the principal focal lengths as in the case of the Ramsden eyepiece.

To locate the equivalent points when the system is negative, mount both lenses on the same side of the holder H, Fig. 117. Place between the source and L_2 an auxiliary positive lens L_3,

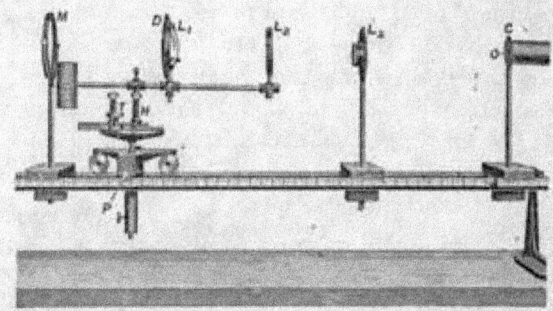

FIG. 117.

and locate the corresponding equivalent point and find the principal focal length as described in Exp. 32.

Make diagrams of the four cases which show the positions of lenses, equivalent points, and principal focal points of both systems.

Exp. 34. Study of the Eye by Means of Kuehne's Model

THEORY OF THE EXPERIMENT. — Optically, the human eye consists of an aperture P (called the pupil) in a diaphragm I

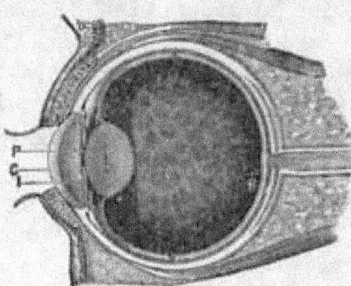

FIG. 118.

(called the iris), a lens L, and a screen R (called the retina). These parts are enclosed in a nearly spherical opaque envelope S provided with a curved transparent round window C (called the cornea). Filling the space between the cornea and the lens is a fluid called the aqueous humor, and between the lens and the retina a semi-fluid substance called the vitreous humor. Accommodation or focalizing on the fixed retina the images of objects at different distances from the eye is effected by a change in the convexity of the lens produced by a muscle around the edge of the lens.

In this experiment, the function of the various optical parts of the eye, and some of the common errors of refraction, are to be studied by means of a model of the eye due to Kuehne. The body of Kuehne's Eye Model, Fig. 119, is a rectangular box with one transparent side and one transparent end. In the front of the box is a tube C which contains a round curved glass corresponding to the cornea. A diaphragm I directly behind C represents the iris, and a convex lens L represents the crystalline lens. The ground glass R represents the retina; the water between the lens and iris represents the aqueous humor; and the water between the lens and ground glass represents the vitreous humor. In front of the cornea is a frame A for the reception of spectacle lenses.

FIG. 119.

MANIPULATION. — Place the regular cornea on the eye and insert the iris with the large pupil just behind it. Fill the model with water to which has been added a few drops of eosin solution to render the path of light through the water more visible. For use as a luminous object place at a distance of about 70 cm. in front of the cornea a box provided with a hole covered with ground glass. An eight-armed cross is a very satisfactory form for the hole. An incandescent lamp in the box renders the object bright and distinct.

(a) *Accommodation*. — Place in the model the lower power lens L so that the pin touches the iris. Move the retina back and forth till a clear image appears. This is the normal position of the retina. Compare the image with the object with respect to size and orientation.

Move the object up and down, and from side to side, noting how the image moves. Substitute for the lens L one of higher power L' and, without disturbing the retina, move the object nearer to or farther from the eye till a clear image appears on the retina. Note what change has been made in the image. Compare this image as to size, brightness, and sharpness with the one formed when the other lens was used. How does the eye accommodate itself for different distances?

(b) *Spherical Aberration.* — With the model as in the previous experiment, remove the iris and note any change in the distinctness of the image. Insert the iris with the small pupil and note any change in distinctness. Focalize until a sharp image is produced and substitute the ring diaphragm I'' for the previously used iris and note any change in the image. Can a sharp image be now produced? Note any change of focus. Explain. Note the path of light in each case.

(c) *Far Sight.* — Replace the object at a distance of 70 cm. from the cornea; use the large pupil and high power lens; and focalize by adjusting the position of the retina. Now move the retina forward 7 cm. and observe that the image on the retina is blurred. Light is now traversing the model as in a far-sighted eye. Find what kind and what power of spectacle lens placed in the holder A will cause the image on the retina to be sharp. Note the path of light through the model with and without the spectacle lens.

(d) *Near Sight.* — Remove the spectacle lens and refocalize by adjusting the position of the retina. Move the retina back 8 cm. from this position. Light is now traversing the model as in a near-sighted eye. Find the spectacle lens which placed in the holder A will cause the image to be sharp. Note the path of light through the model with and without the spectacle lens.

(e) *Corneal Astigmatism.* — Empty the eye and substitute for the regular cornea a cylindrical lens, placing the axis of the lens vertical. Refill the model and place it 70 cm. from the object. Use the low power lens L and no iris. Note the path of light. Observe that there is one position of the retina for which vertical lines are in focus, a different position for which horizontal lines are in focus, and no position for which lines in other directions are in focus. Explain.

Adjust the retina till vertical lines are in focus and find what kind of spectacle lens placed in the frame A will cause the image of all lines of the object to be distinct. Note the power of the spectacle lens and the position of the axis.

Remove the spectacle lens, adjust the retina till horizontal lines are in focus, and find the spectacle lens that will cause all lines of the image to be distinct.

Remove the spectacle lens, rotate the cornea 45°, and repeat the procedure of the preceding part of this article.

(f) *Vision without a Lens.* — Remove the lens L and the spectacle lens, and replace the iris and regular cornea. Determine whether a distinct image is formed for any position of the retina. Find what spectacle lens will cause the image to be most distinct when the retina is in its normal position.

Arrange all observations in a table and state clearly the conclusions to be drawn from each part of the experiment.

Exp. 35. A Study of Telescopes

THEORY OF THE EXPERIMENT. — A telescope is an optical instrument designed to increase the magnitude and the resolution of the retinal images of distant objects. The essential parts of a telescope are two in number — first, a device for collecting light emitted by the object and bringing it to a sharp focus; second, a device for magnifying the image thereby produced. The first device is called the objective; the second is called the ocular or eyepiece. The objective may be a converging lens or mirror and the ocular may be a converging or a diverging lens or lens system.

To be satisfactory a telescope must produce an image possessing the following qualities: — distinctness, magnitude, brightness, large field of view, freedom from curvature of field and distortion, freedom from false colors.

The degree of resolution and the brightness of the image depend upon the diameters of the objective and ocular; the magnitude of the image depends upon the focal lengths of objective and ocular; the field of view depends upon the arrangement of the optical system; the flatness of field and freedom from distortion depend upon the curvature of the optical surfaces; freedom from false colors depends upon the focal lengths and the arrangement of the parts of the optical system as well as upon the material of which the lenses are made.

The Galilean telescope consists of a positive objective and a negative ocular. When focalized for a distant object and most easy vision the distance between the lenses is equal to the difference of the focal lengths of the two lenses. There is no real image, and consequently a cross hair cannot be employed.

The Kepler or simple astronomical telescope consists of a positive objective and a positive eye-lens. For most easy vision the eye-lens is placed at its focal distance from the real image formed by light that has traversed the objective. Cross hairs may be placed in the plane of the image.

The Ramsden eyepiece consists of two plano-convex lenses of the same focal length, with their convex faces toward one another and separated by a distance equal to two-thirds the focal length of either lens. The principal focal planes of the combination are outside the combination at a distance one-fourth of the focal length of either lens measured from the first principal plane of the nearer lens. This eyepiece can therefore be used in place of a simple ocular in any optical instrument with the advantage of a clearer image and greater field of view.

The Huyghens eyepiece used in this experiment consists of two plano-convex lenses, the eye-lens having a focal length equal to one-third that of the field lens. They are separated by a distance equal to the difference of their focal lengths and the convex surface of each is directed toward the objective. When placed in front of an objective at the proper distance to give most easy vision, the aerial object due to light that has traversed the objective is between the lenses of the eyepiece at a distance from the field lens equal to one-half the focal length of the field lens. The real image is formed midway between the two lenses of the ocular. If this ocular has small magnifying power, cross hairs may be placed in the real image.

The Terrestrial Telescope consists of an objective and an erecting eyepiece. The simplest type of erecting eyepiece consists of two convex lenses in combination with a simple positive eye-lens. The two positive erecting lenses may also be added to either a Ramsden or a Huyghens eyepiece. The image of the object, formed by light after traversing the objective, is formed in the focal plane of the first erecting lens. The light leaves the first erecting lens with a plane wave front, and is incident on the second erecting lens placed at any convenient distance from the first. In case the simple one-lens ocular is used, the light from the second erecting lens is brought to a focus in the principal focal plane of the ocular. If the Ramsden or the Huyghens eyepiece is used, the eyepiece must be so placed that the real image shall be formed in the proper position for the particular eyepiece.

The object of this experiment is to assemble lenses so as to form the various standard types of refracting telescope; to place cross hairs where they can be used; to observe the size of the field of view in each type and also the chromatic and spherical aberration.

MANIPULATION. — The apparatus to be used in this experiment consists of a horizontal bench b, Fig. 120, a supplementary

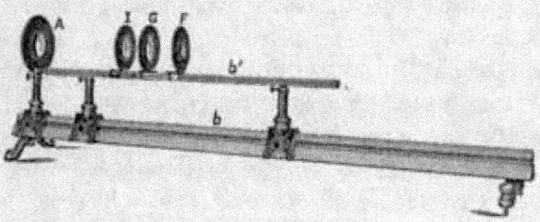

FIG. 120.

bed b', together with various lenses, diaphragms, and cross hairs. The objective can be mounted on the optical bench, and the parts constituting the eyepiece can be mounted on the supplementary bed.

From the collection of lenses of known focal lengths furnished with the equipment, the student will assemble the following telescopes: (a) Galileo's; (b) Simple Astronomical; (c) Astronomical with Ramsden's Eyepiece; (d) Astronomical with Huyghens' Eyepiece; (e) Terrestrial with Huyghens' Eyepiece.

The Galilean Telescope. — Place the corrected lens marked "*A*"
in the support at the end of the optical bench nearer the object to
be observed. By means of a white card, find the position of the
real image. This position can be marked temporarily by means of
cross hairs mounted on the supplementary bed. Place the concave lens *H* nearer the objective than the real image by a distance
equal to the focal length of the concave lens. The telescope
should now be in focus for the eye at rest. Note the size of the
field of view and any indication of chromatic or spherical aberration.

Replace the corrected objective *A* by the uncorrected objective
B and note any change in the aberrations.

*In order not to strain the eyes, always have both eyes open when
looking into any optical instrument.*

For this and each subsequent instrument studied, sketch a
ray diagram in which are indicated the position of lenses, all

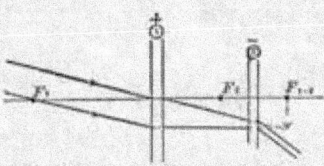

images (real, virtual, and aerial),
and the cross hairs. A ray diagram of the Galilean telescope, but
without distances being indicated,
is given in Fig. 121.

FIG. 121. — Galileo's Telescope.

*The Simple Astronomical Tele-
scope.* — With the lens *A* as ob-
jective, find the position of the real image of the object sighted
upon. Mark this position with the cross hairs. Place the lens

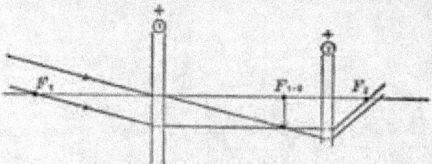

FIG. 122. — Simple Two-lens Astronomical Telescope.

E farther from the objective than this image by a distance equal
to the focal length of *E*. The telescope should now be in focus
for the eye at rest. Make the same observations and diagram
as in the case of the previous instrument.

Replace the corrected objective *A* by the uncorrected objective
B and note any change in the aberrations.

The Astronomical Telescope with Ramsden's Eyepiece. — With
lens *A* as objective substitute for the simple ocular of the previous
instrument a Ramsden eyepiece formed by field lens *G* and eye-
lens *F*, Fig. 123. The distance between the incident principal
plane of the field lens and the real image produced by the light
after traversing the objective should be one-fourth the principal

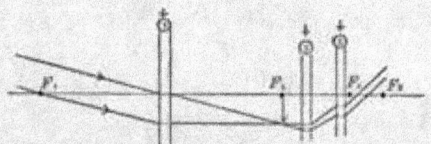

Fig. 123. — Telescope with Ramsden's Eyepiece.

focal length of one of the eyepiece lenses. Place the cross hairs in
the proper position.

Replace the lens *A* by lens *B* and note any tendency to aberration.

Make the same observations and diagram as in the preceding
cases.

The Astronomical Telescope with Huyghens' Eyepiece. — With
lens *A* as objective substitute for the Ramsden eyepiece a Huyghens eyepiece formed of field lens *J* and eye-lens *E*. Place the

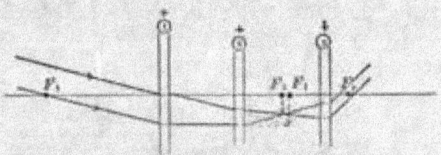

Fig. 124. — Telescope with Huyghens' Eyepiece.

Huyghens eyepiece so that the aerial object, due to light that has
traversed the objective, is between the lenses of the ocular at a distance from the field lens equal to one-half the focal length of this
lens. The insertion of the field lens changes the position of the real
image to a position midway between the lenses. Place cross hairs
at this position. The telescope should now be in focus for the eye
at rest. Since the real image is curved and the cross hairs plane,

parallax may appear in the border of the field. Reduce this by
stops, noting their position.

Replace objective A by B, noting any tendency to aberration.
Make the same observations and diagrams as before.

The Terrestrial Telescope with Huyghens' Eyepiece. — Use lens
A as objective. At a point distant from the real image equal
to its focal length place the first erecting lens C. Place the second
erecting lens D a short distance from the first (a distance equal

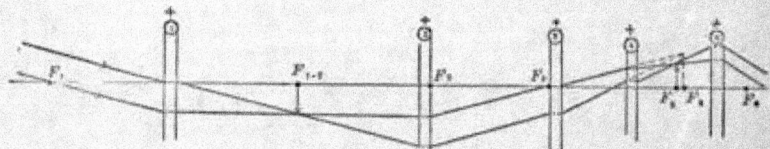

FIG. 125. — Telescope with Huyghens' Eyepiece and Erecting Lenses.

to its focal length provides the greatest correction of aberration),
and locate the real image. Place the Huyghens eyepiece as indicated above with reference to this real image. The telescope is
now in focus for the eye at rest.

Make the same observations and diagram as in the preceding
cases.

Exp. 36. Determination of the Magnifying Power of a Reading Telescope

THEORY OF THE EXPERIMENT. — The magnifying power of a
telescope is the ratio of the angle at the eye subtended by an image,
to the angle at the eye subtended by the object. But since the
distance between the object and objective lens is so nearly equal
to the distance between the object and the eye, the magnifying
power is almost equal to the ratio of the angle subtended at the eye
by the image, to the angle subtended at the objective lens by the
object. Thus, in Fig. 126, the magnifying power

$$M = \frac{\Delta}{\delta}.$$
(177)

The purpose of this experiment is to determine the magnifying power of a reading telescope for various object distances. Two methods are to be employed, — one of considerable precision that involves the use of an accurately divided circular scale and a supplementary telescope, and another, of much less precision, that involves no accessory apparatus.

The Gauss Method. — Light traversing any optical path will retrace the same path if proceeding in the opposite direction. Thus, if in Fig. 126 the object be to the right of the eye-lens, then

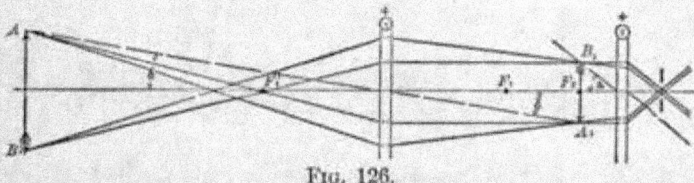

FIG. 126.

light from the object incident upon the eye-lens in the directions represented will form an image at AB. In this case the image is smaller than the object. The diminishing or minifying power of the reversed telescope is then

$$m = \frac{\Delta}{\delta}. \qquad (178)$$

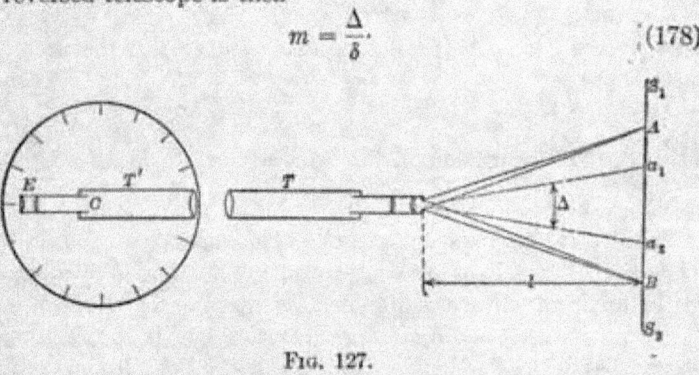

FIG. 127.

That is, the minifying power of a telescope focalized for distant objects equals the magnifying power.

This fact is the basis of the Gauss method for the determination of magnifying power. In this method, the telescope under investigation, T, Figs. 127 and 128, is first focalized on some distant

object and is then pointed away from a horizontal scale S_1S_2. An
auxiliary telescope T', capable of rotation about a vertical axis
through C, is placed with its objective close to that of the telescope
under investigation, and is so adjusted that with the eye at E a
clear image of the scale is seen which does not move with reference
to the cross hairs of T' when the observer's eye is slightly moved.

FIG. 128.

With the telescopes in line as in the figure, the field of view ex-
tends from A to B. If T' be slightly rotated about the vertical
axis through C, its own field of view moves with it, while that of T'
remains fixed. Let the auxiliary telescope T' be rotated to near
the end of the observed field of view. Suppose that the intersection of the cross hairs coincides with the image of some division
a_1 of the horizontal scale when the angular position of the telescope
with reference to the circular scale is ϕ_1. Now let the auxiliary
telescope be rotated to near the other end of the field of view.
Suppose that now the intersection of the cross hairs coincides with
the image of some division a_2 of the horizontal scale when the angular position of the telescope is ϕ_2.

The angle at the eyepiece of the telescope under investigation
between rays from a_1 and a_2 is Δ, and the angle at the objective of
the same telescope between the emerging rays from a_1 and a_2 is
$(\phi_2 - \phi_1)$. Thus, from Fig. 127,

$$M\left[= m = \frac{\Delta}{\delta}\right] = \frac{2\tan^{-}\left(\dfrac{a_1 - a_2}{2\,l}\right)}{\phi_2 - \phi_1}. \tag{179}$$

Approximate Method. — Since the tangents of small angles are approximately proportional to the angles, the magnifying power of an optical instrument approximately equals the ratio of the size of the image seen when looking through the instrument to the size of the image seen by the unaided eye. Thus an approximate value of the magnifying power of a telescope can be obtained by comparing the size of the image of an object seen by one eye placed at the eyepiece, with the size of the image of the same object seen by the other eye without the telescope.

MANIPULATION. — *The Gauss Method.* — Focalize the telescope under investigation on a distant object. Arrange this telescope and an accessory telescope as shown in Fig. 128, and determine the magnifying power as above described. This is the magnifying power for distant objects.

Approximate Method. — With the telescope five meters from a scale adjust the eyepiece till the image of the scale seen by the eye at the telescope coincides with the position of the scale as seen by the other eye. When this adjustment is effected, there will be no relative motion between the two images as the head is moved slightly from side to side. The first time the attempt is made to view simultaneously a different image with each eye, some difficulty will be experienced in preventing the attention from being fixed on one image. But usually a few minutes training will suffice to develop the necessary control of the attention. Since the two eyes are on a horizontal line, the outside of the telescope will be least obtrusive if the scale be vertical. When the telescope is in proper adjustment, the two images of the scale will appear side by side, the one seen by the unaided eye smaller than the other, something as shown in Fig. 129. In this figure, three divisions of the scale as seen by the unaided eye correspond to one division as seen by the eye at the telescope. Thus for this particular object distance, the magnifying power is three.

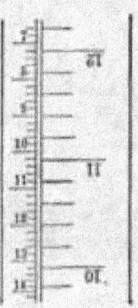

FIG. 129.

In the same manner determine the magnifying power for object distances of three and for 1.25 meters.

Exp. 37. A Study of the Resolving Power of a Telescope

Theory of the Experiment. — Read Arts. 36 and 37. When the images of two object points can be distinguished as separate images, the two point sources are said to be resolved. Whether two points are barely resolved, or not, depends upon one's criterion of resolution. Two observers will usually differ as to the exact amount of separation necessary to constitute the limit of resolution. But for purposes of analysis and the assigning of a numerical value to the resolving power of lenses it is customary to arbitrarily postulate that two object points are at their limit of resolution when the centers of the images are separated by a distance equal to the radius of the central diffraction disk of one of them. Using this convention, we find that the resolving power of a lens covered with a diaphragm containing a narrow slit of width a is, (125),

$$\left(\frac{1}{\theta}\right)_{s} \doteq \frac{a}{\lambda}, \tag{180}$$

and that the resolving power of a lens with a circular aperture of diameter a is, (128),

$$\left(\frac{1}{\theta}\right)_{c} \doteq \frac{a}{1.22\,\lambda}. \tag{181}$$

The resolving power of a telescope equals that of the lens which has the smallest resolving power. Since the resolving power of a lens is proportional to the ratio of the diameter of the transmitted beam of light to the focal length of the lens, Art. 36, the resolving power of the short focus lenses composing the ocular of telescopes is almost always larger than that of the objective. Whence, the resolving power of a telescope is usually determined by that of the objective.

The object of this experiment is to change the aperture of the objective of a telescope directed toward an object consisting of sharp parallel lines separated by a known distance till, from an inspection of the image, the observer judges that the limit of resolution is attained; from the distance u between the object and objective, together with the distance D between the parallel object lines, the practical resolving power of the telescope is to be computed by means of (121),

$$\frac{1}{\theta} = \frac{u}{D}.$$ (182)

This value of the practical resolving power is then to be compared with the theoretical resolving power of the objective obtained by means of (180) or (181) as the case may be.

MANIPULATION. — For this experiment the object consists of a series of equally spaced parallel lines ruled on a glass plate and illumined by transmitted light from a sodium burner. Such

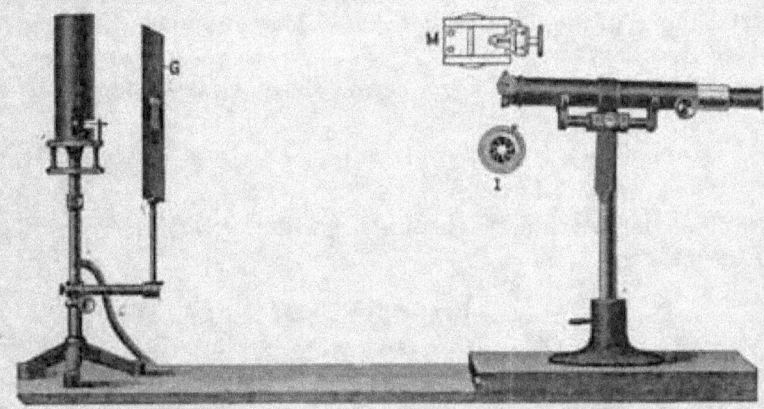

FIG. 130.

gratings may be made by scratching through the black coating of an exposed and developed photographic plate. In Fig. 130 the object is shown at the center of the frame G.

To fit over the objective end of the telescope under test are provided two diaphragms M and I. The first is provided with a slit the jaws of which are operated by a micrometer screw. The second is similar to the iris diaphragms used on cameras.

In the frame in front of the sodium flame place the grating with the rulings vertical. Place in front of the telescope objective the micrometer slit with the slit vertical. With one eye at the telescope ocular, gradually open the micrometer slit and note the changes in the appearance of the image. In the report, explain the appearances observed.

When near the limit of resolution, the bright image lines become broader and nebulous at the edges. Beyond the limit of resolution the field of view becomes without structure and evenly bright. This appearance is different than that produced by longitudinal spherical aberration. When a lens has spherical aberration, the spaces between sharp image lines will be hazy but uniform.

With the telescope placed about 125 cm. from the object, and with the micrometer slit wide open, focalize the telescope. Now close the micrometer slit until the images of the lines of the object can barely be distinguished as separate. Measure the distance u from object to objective. Read the width a of the micrometer slit. The distance D between adjacent object lines is marked on the glass plate. By means of (182) compute the practical resolving power of the objective. Assuming the wave-length of sodium light to be 0.0000589 cm., compute by means of (180) the theoretical resolving power of the lens with a slit aperture.

Substitute the iris diaphragm for the slit diaphragm and find the practical as well as the theoretical resolving power of the lens for a circular aperture. The diameter of the aperture in the iris diaphragm when the object is just resolved can be measured with sufficient precision by means of a steel scale.

Repeat readings with the slit diaphragm and with the iris diaphragm for object distances of about 200 cm. and of 250 cm.

Exp. 38. Determination of the Refractive Index of a Substance in the Form of a Prism

THEORY OF THE EXPERIMENT. — Read Arts. 38–40.

MANIPULATION. — In this experiment will be used a spectrometer, Fig. 131, consisting of a collimator with the slit S directed toward a sodium flame not shown in the figure, and a telescope T mounted so that it can be rotated about a vertical axis through the center of the horizontal divided circle A. The prism P under investigation is placed on a table which is capable of rotation about the same axis. The position of the telescope relative to the divided circle can be read by means of two verniers 180° apart. Some spectrometers are also provided with a divided circle B for reading the position of the collimator.

The manipulation of this experiment includes the leveling of the collimator, telescope, prism and divided circular table, the adjustment of the optical system of the telescope and collimator, the

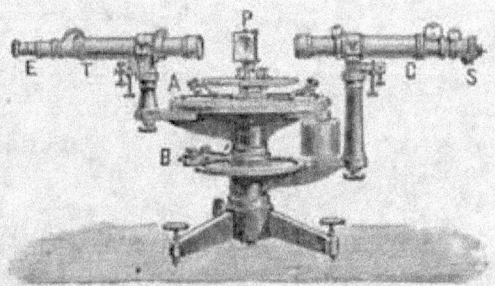

Fig. 131.

measurement of the refracting angle of the prism, and the measurement of the minimum angle of deviation of the prism when sodium light is used. The data will then be at hand for determining by (137) the refractive index of the material composing the prism when transmitting sodium light.

Remove the telescope, point it toward a distant object, and adjust the position of the objective and the ocular till the image of the distant object is in the plane of the cross hairs. When properly adjusted, there is no motion of the image relative to the cross hairs on moving the eye from side to side in front of the ocular.

Place a lamp in front of the collimator slit, replace the telescope, and put the telescope and collimator in line. Without changing the focus of the telescope, focalize the collimator till the image of the slit is in the plane of the cross hairs.

By the aid of spirit levels and adjusting screws, make level the table, telescope, and collimator.

Place the prism on the table of the instrument with the refracting edge at the center of the table and directed toward the collimator. Turn the collimator slit till it is horizontal. Move the

telescope until an image of the slit reflected from one face of the
prism is seen in the eyepiece. Adjust the level of the prism till
the image is at the intersection of the cross hairs. Now move the
telescope till the image of the slit reflected from the other face of
the prism is seen in the eyepiece, and adjust the level of the prism
as before. Turn back the telescope and repeat the adjustment of
the prism till both reflected images come in the intersection of the
cross hairs.

The instrument is now in adjustment. The adjustment must
not be altered throughout the subsequent experiment.

To find the refracting angle of the prism, set the collimator slit
vertical and turn the telescope till the image of the slit reflected
from one side of the prism coincides with the intersection of the
cross hairs. Denote by $T_1°$ the angular position of the telescope
relative to the divided circle. Turn the
telescope till the image of the slit reflected
from the other face of the prism coincides
with the intersection of the cross hairs. Call
the present scale reading $T_2°$. Then from
Fig. 132, the angle through which the telescope has been turned is

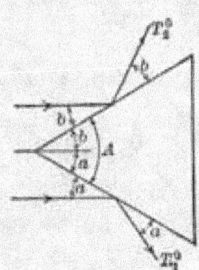

$$T_1° - T_2° \ [= a + A + b] = 2A. \quad (183)$$

Fig. 132.

In case the angular position of the telescope
is given by two verniers 180° apart reading
T' and T'', respectively, then the true readings to be used in the
above equation are

$$T_1° = \tfrac{1}{2}[T_1' + (T_1'' - 180)] \quad \text{and} \quad T_2° = \tfrac{1}{2}[T_2' + (T_2'' - 180)].$$

To find the angle of minimum deviation of the prism, move the
telescope till light traversing the prism forms a spectrum in the
eyepiece. Now rotate the prism, keeping the sodium lines on the
cross hairs by moving the telescope, till the deviation is increased
when the prism is rotated in either direction. The prism is now
set at minimum deviation for sodium light. Note the scale reading D_1. Again rotate the prism and the telescope till another
spectrum is observed on the other side of the axis of the collimator.

Find, as before, the position of minimum deviation. Note the scale reading D_2. The angle between the readings D_1 and D_2 is twice the angle of minimum deviation δ.

Substituting these values of A and δ in (137), compute the index of refraction of the material composing the prism for sodium light.

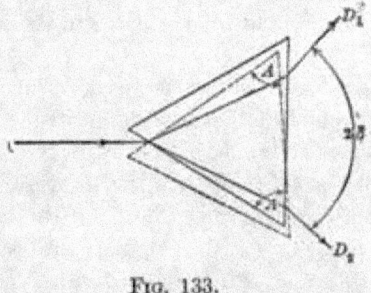

FIG. 133.

Exp. 39. Determination of the Refractive Index of a Liquid

THEORY OF THE EXPERIMENT. — Read Arts. 38 and 41. The object of this experiment is to determine the refractive index of a specimen of a given liquid by means of two or more of the refractometers described in Arts. 43–49, and to compare the advantages and disadvantages of the methods employed.

MANIPULATION. — The parts of the apparatus in contact with the specimen must be scrupulously cleaned before and after the experiment. The methods of cleaning, as well as the other manipulative details of the experiment, are left to the ingenuity of the student. With each instrument, light of the same wave-length should be used.

Exp. 40. Determination of Specific Refractivities and the Composition of a Mixture

THEORY OF THE EXPERIMENT. — Read Art. 42. The object of this experiment is to first determine the specific refractivity of a mixture of two substances and of a specimen of each component, and then by means of (145) to compute the proportion of the components in the mixture. This method for determining composition can be employed only when there is a considerable difference between the specific refractivities of the components. For the following pairs of liquids the method is available: acetone and toluene, acetone and carbolic acid, acetic acid and benzene, ethyl alcohol and benzene, ethyl alcohol and carbolic acid. The method is also available for certain soluble salts in solution, for example, potassium chloride in water.

MANIPULATION. — Make a mixture, or solution, of the two substances, by weight, in known proportion. Find the density, and also the refractive index, of a sample of each ingredient and of the mixture, at room temperature. For finding the densities, the pyknometer shown in Fig. 24 is well suited. For finding the refractive indices, any of the refractometers described in Arts. 43–49 can be employed.

From these data compute the Lorenz specific refractivity of each component and of the mixture. Using these values, compute the proportion of the constituents by (145). Compare this computed value of the composition with the known composition.

Exp. 41. Study of Spectra

THEORY OF THE EXPERIMENT. — Read Arts. 52–54. The fact that the bright line spectrum of any element is different from that of all other elements, together with the fact that the spectrum of a mixture consists in the spectra of the components, is the basis of an important method of qualitative chemical analysis. A continuous spectrum means that the incandescent substance is either a solid or a liquid. An absorption spectrum implies an incandescent solid or liquid together with a layer of cooler absorptive material which may be solid, liquid, or gas. The object of this experiment is to produce and study examples of bright line, continuous, and absorption spectra.

Different means must be employed to vaporize different classes of substances. The alkaline earths are readily transformed into incandescent vapor by the heat of a Bunsen burner. Gases are easily rendered incandescent by a current of electricity. Some metals require the heat of an electric arc, while others are more readily vaporized by means of an electric spark discharge. In this experiment are to be examined, (*a*) the spectra of two or more alkaline earths, (*b*) the spectrum of an incandescent permanent gas, (*c*) the continuous spectrum from burning illuminating gas, (*d*) the absorption spectrum of a colored glass or solution.

MANIPULATION. — A convenient arrangement of apparatus for this experiment is illustrated in Fig. 134. On a board are mounted a direct vision spectroscope with an incandescent lamp *l* for illuminating the comparison scale, together with a device *V* for holding a

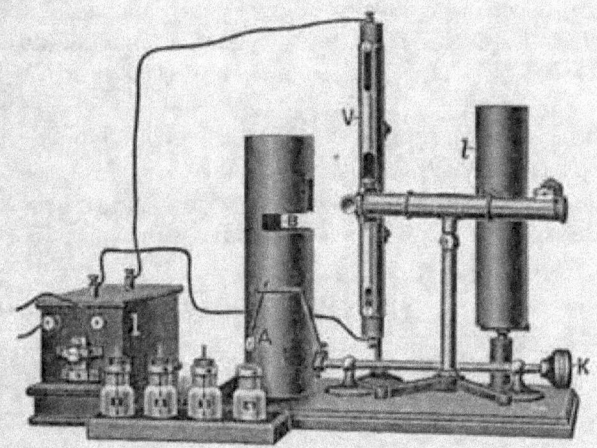

FIG. 134.

gas tube and an induction coil to render the gas luminous, as well as a Bunsen burner *B* for vaporizing the specimens shown in the bottles. By turning the knob *K* a specimen *A* whose spectrum is to be studied may be brought into the Bunsen flame or into the proper specimen bottle without moving the eye from the ocular of the spectroscope.

Light the gas, close the air vent to the Bunsen burner and observe the spectrum.

Turn on the current in the lamp that illumines the comparison

scale. Place before the slit a piece of colored glass and observe
the spectrum. Note the scale divisions at the boundaries of the
dark bands. Do the same with a vial of cobalt chloride solution
acidulated with hydrochloric acid.

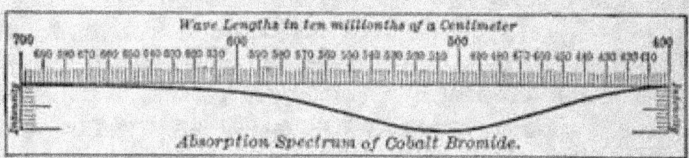

Fig. 135.

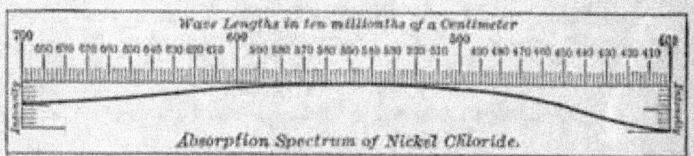

Fig. 136.

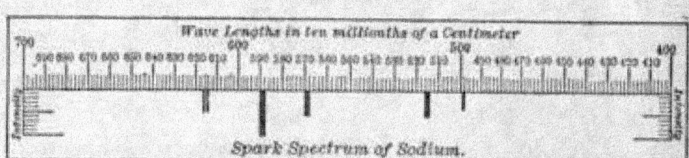

Fig. 137.

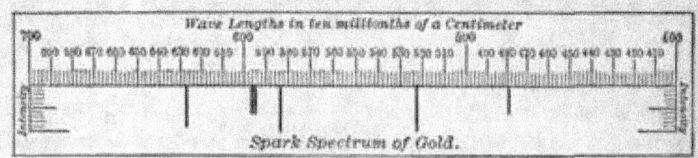

Fig. 138.

Open the air vent until the gas burns with a blue flame. Clamp
the wire contained in the NaCl bottle into the free end of the rotating arm that is operated by the knob K. Twist the knob till the
asbestos string on the end of the specimen wire is brought into the
flame. Observe the spectrum. Focalize the spectroscope till
the bright sodium line is sharp. With an instrument of greater

resolving power the sodium line would be double.   Note the scale position of the line.   Remove the specimen wire and replace it in the proper bottle.

In the same manner examine the spectrum of each of the other solutions furnished, and note the scale positions of all lines.   Be careful that each specimen wire is replaced in the proper bottle.

Now turn the totally reflecting prism into place in front of the slit, and turn off the gas.   Connect the primary of the induction coil, I, to a proper source of current, and the secondary to the terminals of the device V, that contains the tube of gas to be studied. Note the scale position of each spectral line.

Make a map similar to Figs. 135–138 of the spectrum of each substance studied.   In the case of a bright line spectrum, the positions of the lines in the map correspond to either the positions of the spectral lines on the scale of the instrument, or to the wave-lengths of the spectral lines.   The length of a line in the map indicates the relative brightness of the spectral line which it represents. In the case of an absorption spectrum, the ordinates of the absorption curve represent the relative brightness of the spectrum at various wave-lengths.

Exp. 42. Calibration of a Prism Spectroscope by Spectral Lines of known Wave-Lengths

THEORY OF THE EXPERIMENT. — Read Art. 53.   Spectroscopes in which the prism and lenses are fixed with reference to one another are usually provided with a device by means of which a bright image of a linear scale is formed in the focal plane of the eyepiece.   The positions of spectral lines can be thereby indicated in terms of this arbitrary scale.   In the case of spectroscopes in which the prism and telescope can be rotated, the position of spectral lines can be described in terms of the angular position of the telescope relative to the collimator when the prism is in the position of minimum deviation for light of a given wave-length.

Instead of specifying a spectral line in terms of the constants of a particular instrument, it is usually preferable to specify it by its wave-length.   By noting the position, relative to the spectroscope scale, of several spectral lines of known wave-lengths, a curve can be constructed that shows the relation between wave-lengths and scale divisions.

Spectral lines of known wave-lengths can be obtained from rarefied gases rendered luminous by an induction coil, or from chlorides of the alkaline metals and earths vaporized in the flame of a Bunsen burner. For purposes of calibration it will be found convenient to use the vacuum tube spectrum of helium, and the flame spectra obtained from the chlorides of the alkaline metals and earths. The principal visible lines of these substances have the following wave-lengths, expressed in Ångström units.*

> *Helium* — 4472, 4713, 4922, 5016, 5876, 6678;
> *Barium* — 5536;
> *Cadmium* — 4678, 4800, 5086;
> *Calcium* — 5817, 5934, 6266;
> *Lithium* — 6104, 6708;
> *Potassium* — 4044, 7665, 7699;
> *Strontium* — 6351, 6599, 6730;
> *Sodium* — 5893;
> *Thalium* — 5350.

MANIPULATION. — In case the instrument is with fixed parts, place a lamp in front of the side tube containing the glass scale, and a sodium flame in front of the slit. Focalize the eyepiece so that the image of the scale and the sodium line do not move relative to one another when the eye is moved from one side to the other in front of the eyepiece.

In case the instrument is one with a divided circular table and a movable prism and telescope, make the adjustments detailed in Exp. 38.

If a helium tube be available, place it in front of the slit and take scale readings of the various lines. With the instrument having all the parts fixed, the positions of the various lines are read directly from the eyepiece scale. In the case of the instrument with movable telescope, rotate the telescope till the intersection of the

* The Ångström unit is 10^{-10} meters $= 0.000\,000\,1$ millimeter $= 0.0001$ micron $= 0.1\ \mu\mu$.

cross hairs is on the middle of a spectral line, and read the position
of the telescope on the circular table.   Do the same with the other
spectral lines.

If a helium tube is not available, one can employ concentrated
solutions of some of the alkaline metals or earths, acidulated with
hydrochloric acid.   The solutions can be conveniently introduced
into the Bunsen flame by means of a wick of asbestos in the manner
illustrated in Fig. 134.

Plot a curve coördinating scale readings and wave-lengths.   This
is the calibration curve required.

Exp. 43. Construction of a Simple Plane Grating Spectrometer and the Determination of Wave-Lengths of Light

THEORY OF THE EXPERIMENT. — Read Arts. 50 and 53.   In
this experiment a transmission spectroscope is to be built up from
a grating, lenses, and rods, and by means of it are to be determined
the wave-lengths of light emitted by a given incandescent vapor.
The essential parts of a simple transmission grating spectroscope

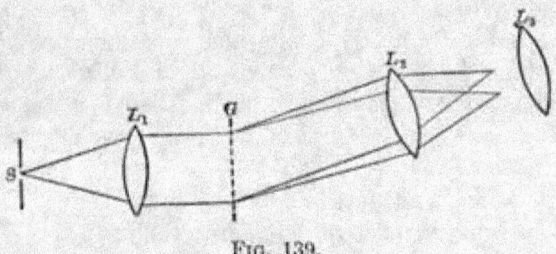

FIG. 139.

are a slit S, Figs. 139 and 140, a lens L_1 for rendering the light from
the slit parallel, the grating G, a lens L_2 for focalizing the light from
the grating, a pair of cross hairs D in the focal plane of the objective L, an eyepiece L_3 for magnifying the images of the slit, and a
divided circular scale for measuring deviations.   The slit and lens
L_1 constitute a system called a collimator.   The objective L_2, eyepiece L_3, and cross hairs D in the image plane constitute a tele-

scope. Attached to the telescope is a pointer so that the position of the telescope can be read on the divided circle.

MANIPULATION. — Set up the collimator by mounting on one of the horizontal rods the lens L_1, the slit S, and the light source. The distance between the lens and slit should be such that the light on emerging from the lens shall be parallel. To make this

FIG. 140.

adjustment, light the lamp; and by means of a small hand mirror reflect the light emerging from the lens back through the same lens, and observe the image of the slit formed beside the slit. Adjust the position of the lens till this image of the slit is well defined. The slit is now at the principal focus of the lens, and light emerging from the lens is parallel.

Place the collimator so that its optic axis intersects the axis of the divided circle at right angles. On the other horizontal rod, and in line with the collimator, mount the lenses and cross hairs constituting the telescope, and adjust till the image of the slit is sharply defined in the plane of the cross hairs.

Place the grating at the center of the divided circle with the ruled surface in the plane of the axis of the divided circle and the rulings parallel to this axis. The slit should also be parallel to this axis. The adjustment of the plane of a transmission grating normal to the direction of the incident light can be tested as follows: With a sodium flame in front of the slit, read the positions of the pointer when the cross hairs coincide (a) with the central image; (b) with the first order image to the right; (c) with the first order image to the left. If the grating be normal to the incident light, then the angle between the central image and the first order image to the right will equal the angle between the central image and the first order image to the left.

After the instrument is in adjustment, measure the deviation of the sodium lines from the central image and by means of (152) compute their wave-lengths.

Exp. 44. Determination of Wave-Lengths of Light by Means of a Steinheil Spectrograph

THEORY OF THE EXPERIMENT. — Read Art. 50. The Steinheil spectograph is designed for the convenient application of (153). Light from the fixed collimator C, Fig. 141, after being totally reflected by the fixed prism P, and being diffracted by the plane grating AB, enters the fixed telescope T. The grating is capable of rotation. The angle through which the grating is rotated is indicated by an alidade (*i.e.,* pointer) and circular scale.

Equation (153) may be written

$$\lambda = \frac{k}{n}\sin w, \qquad (184)$$

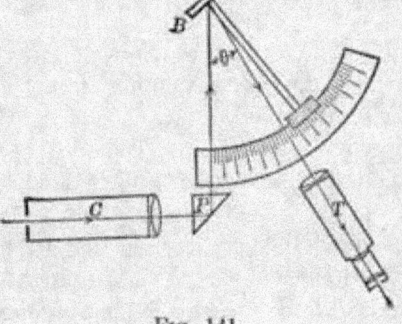

FIG. 141.

where k represents the constant quantity $2\,b\cos\dfrac{\theta}{2}$ and w represents $\frac{1}{2}\,\phi$. The value of the constant k is marked by the maker on the instrument. It can be readily determined from a setting on a spectrum line of known wave-length.

The instrument is provided with a camera which can be substituted for the telescope. It is for this reason that the instrument is called a spectrograph.

MANIPULATION. — Turn the grating till the reflected image of the illuminated slit is on the cross hairs of the telescope and note the scale reading. Then turn the grating till the desired spectral line is on the cross hairs and note the scale reading. The angle between these two readings is the value of w in (184).

In the same manner find the wave-lengths of all the lines of the spectrum under investigation.

Exp. 45. Determination of Wave-Lengths of Light by Means of a Concave Grating

THEORY OF THE EXPERIMENT. — Read Arts. 51 and 53. In Art. 51 it has been shown that with an illuminated slit, concave grating and image of the slit on the circumference of a circle having as a diameter the radius of curvature of the grating, wave-lengths of the light illuminating the slit can be readily determined. And that if, in addition, the position of the slit on the circle be such that the grating and the image of the slit are at the ends of the diameter of this circle, the wave-length of the light of a spectrum line at the center of curvature of the grating is, (155),

$$\lambda_c = \frac{b \sin i}{n},$$
(185)

where n is the order of the spectrum, b is the distance between two consecutive grating lines, and i is the angle at the grating between lines from the slit and from the image of the slit.

This arrangement is very conveniently realized in the Rowland mounting. Since the grating and the image are at the ends of a diameter of the circle, the triangle CSY, Fig. 88, must at all times be right-angled at the slit. In Rowland's mounting there are two horizontal rails at right angles to one another, on each of which is a two-wheeled truck. These trucks are pivoted to the ends of a rigid rod which has a length equal to the radius of curvature of the grating. The pins connecting the trucks to the ends of the rod are directly above the centers of the tracks. The fixed slit S, Fig. 142, is mounted at the intersection of the rails. At one end of the movable rod is mounted the grating G, and at the other end is mounted a photographic plate or ground glass C for the reception of the spectrum.

The circle passing through C, S, and G moves in space as the

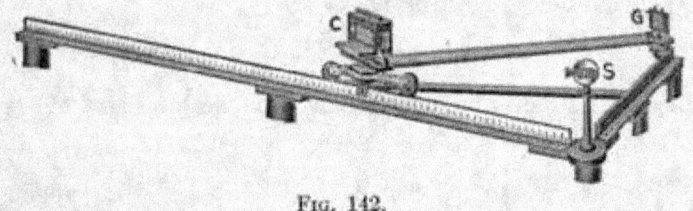

Fig. 142.

rod CG is moved. As the rod of length r is moved, the other sides of the right-angled triangle vary in length. In Fig. 143,

$$\sin i = \frac{l}{r \ (\text{a constant})}.$$

From (185),

$$\sin i = \frac{\lambda_c}{\left(\frac{b}{n}\right) (\text{a constant})}.$$

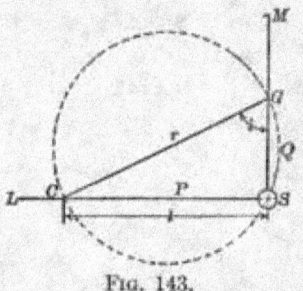

Fig. 143.

Therefore, for a spectrum of a given order, $l \infty \lambda_c$. Consequently the rail that supports the camera can be uniformly divided so as to indicate wave-lengths directly. By bending the photographic plate into an arc of a circle having for a diameter the radius of curvature of the grating surface, all parts of the spectrum on the plate will be in focus at the same time.

Fig. 144.

It is common practice to photograph on the same plate with the spectrum under investigation a comparison spectrum whose lines are of known wave-lengths. The spectrum of the sun is usually employed for this purpose. A convenient device for arranging two spectra side by side on the same plate consists in a shutter having a long narrow opening, Fig. 144, placed near the photo-

graphic plate and toward the grating. The thickness of the
shutter equals the width of the opening. When the plane of
the shutter is vertical, as in the figure, the part of the photographic
plate behind the aperture is exposed to the spectrum of the source
in front of the slit. On turning the plane of the shutter into the
horizontal position, the part of the plate just exposed is now
shielded, and a strip on either side is exposed to the spectrum of
the source now in front of the slit.

Since the wave-lengths of the solar spectral lines have been
carefully determined and tabulated, and since with the Rowland
mounting the difference between the wave-lengths of two lines is
proportional to the linear distance between them, the wave-length
of any line of the spectrum under investigation can be quickly
determined in terms of known lines of the comparison spectrum.

The objects of this experiment are to test the adjustments of a
Rowland concave grating mount, and to photograph a bright line
spectrum and also a comparison spectrum on the same photographic
plate.

MANIPULATION. — Before using the apparatus for determining
wave-lengths, the following adjustments should be verified.

(a) *Perpendicularity of the Tracks.* — Hang plumb bobs over
the centers of the tracks near the ends L and M, Fig. 143, and also
over the centers at convenient points P and Q. The point of
intersection of the centers of the two tracks can now be located
by placing a fifth plumb bob at such a point S that it is at the
same time in line with PL and with QM.

From the point of intersection lay off along the center of each
track by means of a steel tape any convenient distance. Measure
the hypotenuse length from the end points of these distances.
If the tracks are perpendicular to one another, the square of the
hypotenuse equals the sum of the squares of the other two
distances.

(b) *Horizontality of the Tracks.* — Test with a spirit level.

(c) *Position of the Slit.* — Place the slit in line with the plumb
bob which indicates the point of intersection of the tracks.

(d) *Position of the Grating.* — By sighting along the plumb
bobs SGM see that the center of the reflecting surface of the

grating is over the center of the track. To adjust the grating face perpendicular to the rod GC, darken the room, stand a candle near C, and move the head to the right and left, up and down, till the image is found. This should coincide in position with the candle. If it does not, let another observer adjust the screws holding the grating in its mounting till the grating is pointing in about the proper direction. The final adjustment of the grating is made by setting the candle behind a vertical slit cut in cardboard placed at C, and tilting the grating till the slit and its image coincide.

(e) *The Distance between the Grating and the Camera.* — If this distance equals the radius of curvature of the grating surface, a spectrum of a source illuminating the slit S will be in focus over the entire length of a photographic plate at C bent onto an arc of the circle SCG. If the spectrum is not sharp over the entire plate, the distance between the camera and grating must be adjusted till it is.

(f) *Verticality of the Grating Rulings.* — Test and, if necessary, adjust by reference to a plumb bob hung close to the reflecting surface of the grating.

(g) *Position of the Source.* — The grating must be fully and evenly illumined by a beam of light parallel to the track which supports the grating. The slit is illumined by an image of the source produced by the aid of a lens. Darken the room and adjust the position of the source or lens till the shadow of the grating is in the center of the circle of light, and the line joining the center of the slit and the center of the shadow of the grating is parallel to the track.

Place in front of the slit a glass tube containing some known pure gas under diminished pressure. Render the gas brightly luminous by means of a small induction coil. Before introducing a photographic plate into the camera, observe the ground glass while the camera is moved away from the slit and note, (a) that spectra of the first, second, third, etc., orders sweep in succession over the ground glass; (b) that the spectrum of the first order is distinct, but that in the case of higher orders, the red end of one spectrum overlaps the blue of the next higher order; (c) that the blue end of any spectrum passes across the ground glass before the red end. With the camera stationary at any position, note, (a) that the red end of any spectrum is nearer the slit than is the blue end; (b) that a spectrum of higher order is nearer the slit than one of lower order. Show that these facts can be inferred from the equations of Art. 51.

To vaporize a metal or other refractory substance, a direct current arc is usually employed. The two carbons should be vertical and the positive one below. A small piece of the specimen is placed in a shallow cavity in the end of the lower carbon. The length of the arc should be so great that when the image is projected on the slit, the images of the hot carbons can be intercepted without preventing the passage through the slit of the light due to the vaporized specimen.

Using the shutter, Fig. 144, expose the middle of the photographic plate to the spectrum of the specimen, and then expose the portions on either side to the comparison spectrum. The duration of exposure depends so much upon the brightness of the source and the part of the spectrum being photographed, that experience can be the only guide. For the same plate and region of the spectrum, the solar spectrum may require one-half second, while the iron arc spectrum may require thirty seconds and the iron spark spectrum twenty minutes.

The spectrum of helium or other gas is convenient for comparison, though the time of exposure will be long. After one has had some experience, the arc spectrum of iron or the solar spectrum can be used. These have so many lines, however, that considerable experience is necessary to identify them.

Knowing, from tables or maps, the wave-lengths of lines of the comparison spectrum, the wave-lengths of the lines of the other spectrum can be obtained by measuring the linear distances between the various lines on the developed photographic plate. These distances are measured by means of a microscope that can be moved by means of a long micrometer screw from one end of the negative to the other.

Exp. 46. Determination of the Concentration of a Solution by Means of a Martens-Koenig Spectrophotometer

THEORY OF THE EXPERIMENT. — Read Arts. 55, 56, and 59. With the Nicols prism of the spectrophotometer in such a position that one-half of the field of view is dark, suppose absorption cells of thickness n_1 and n_2 filled with a specimen of the solvent used in the solution under test to be placed in front of the slits a and b. Suppose that to make the two halves of the field of view equally bright the Nicol must be rotated through an angle θ. Then, the ratio of the illuminations of the slits by light of a given frequency is, (164),

$$\frac{I_a}{I_b} = \tan^2 \theta. \tag{186}$$

Substituting the solution under test for the pure solvent, let θ' be the angle through which the Nicol must be rotated in order to produce a uniform field. Representing the illuminations of the slits a and b when the incident light has traversed the solution by I_a' and I_b', respectively, we have

$$\frac{I_a'}{I_b'} = \tan^2 \theta'. \tag{187}$$

From (158) we have

$$I_a' = I_a (10)^{-En_1} \quad \text{and} \quad I_b' = I_b (10)^{-En_2},$$

where 10 is the base of the ordinary logarithms and E represents the extinction coefficient of the solution.

Dividing each member of (186) by the corresponding member of (187), and substituting in the resulting equation the above values of I_a' and I_b', we find

$$\frac{\tan^2 \theta}{\tan^2 \theta'} = (10)^{E(n_1 - n_2)},$$

or $\qquad 2 \log \tan \theta - 2 \log \tan \theta' = E(n_1 - n_2)$,

whence,

$$E = \frac{2 (\log \tan \theta - \log \tan \theta')}{n_1 - n_2} \tag{188}$$

Graphical Method. — Knowing the concentration and the extinction coefficient of one solution together with the extinction coefficient of the same solution of unknown concentration, the concentration of the latter can be computed by means of (159). However, when many determinations of concentration of the same solution are to be made the following graphical method is much more expeditious.

Since θ refers to the pure solvent, $\log \tan \theta$ is a constant. The quantity $(n_1 - n_2)$ is also constant. E and $\log \tan \theta'$ are variables. We thus see that the above equation represents a straight line coördinating E and $\log \tan \theta'$.

Since the extinction coefficient varies directly with the concentration of a solution so long as there is no chemical or physical change, it follows from the above that the curve coördinating concentration and $\log \tan \theta'$ is also a straight line. Thus the concentration of a solution of an unhydrolyzed and completely ionized solute can be determined spectrophotometrically. If, however, the solute is partly ionized or partly hydrolyzed, the method is unavailable. The method is applicable whenever the curve coördinating known concentrations and observed values of $\log \tan \theta'$ is a straight line. The concentrations corresponding to the ends of the straight portion of this curve are the limits within which the method can be used.

From the straight portion of the curve coördinating concentration and $\log \tan \theta'$ another curve can be constructed coördinating concentration and θ' within the determined limits of concentration. When this last curve has been once made for a solution of any given solute in a given solvent, the unknown concentration of any specimen of the same solution can be quickly determined if this concentration is within the prescribed limits.

The particular part of the spectrum at which settings should be made depends upon two factors. If, when the two halves of the field of view are in balance, θ' is about $45°$, a given error in reading θ' will cause a smaller error in the calculated value of E than if θ' were small. For example, an error in θ' of $0°.1$ introduces in $\log \tan \theta'$, and consequently in the concentration, an error twice as great when θ' is $15°$ as when θ' is $45°$ On the other hand, at small angles, it is possible to make more accurate judgments of equality of brightness than at angles near 45°. Between these opposing limitations to precision, it is best to compromise by arranging that the balance will occur at angles between 25° and 40°. Again, the absorption at one part of the spectrum may be altered much more by a given change of concentration of the solution than will the absorption at a different part of the spectrum. This alteration in the ratio of the change of absorption to concentration can be determined only by observations at various parts of the spectrum. For example, in the case of two different concentrations of a given solution, the values of θ' differed by 3° when settings were made in the red portion of the spectrum, and 15° when settings were made in the green. Obviously, a given error in θ' in the first case would introduce a far greater change in the calculated concentration than the same error in the second case. In brief, that part of the absorption spectrum should be selected in which the ratio of change of absorption to concentration is the greatest possible consistent with the requirement that θ' shall be between 25° and 40°. For specimens of the same solution differing widely in concentration, the part of the spectrum at which most sensitive readings can be made will not be the same.

The object of the present experiment is to construct the curve coördinating concentrations of a given solution and the corresponding instrumental readings θ'.

MANIPULATION. — The manipulation of this experiment may be divided into the following three steps: (a) The determination of the portion of the spectrum best suited for work with the particular solution being studied; (b) determination of the limits of concentration between which the method is available; (c) construction of the required curve coördinating concentrations and instrumental readings.

FIG. 145.

Two layers of a solution of the same concentration and different thickness are conveniently obtained by the use of the Schulz absorption cell, Fig. 145, consisting of a cell 11 mm. thick in which

is a glass block 10 mm. thick, so arranged that one pencil of
light after traversing 11 mm. of solution enters the slit a, Fig. 146,
while another pencil after traversing 1 mm. of solution and the
"Schulz body" S, enters the slit b. Thus, the difference of
thickness of solution traversed by the two pencils is one centimeter.
The two pencils of light of equal intensity are produced by a

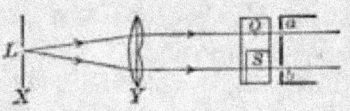

FIG. 146.

triple lens Y. The aperture in
the diaphragm X is covered with
ground glass.

Using a suitable known concentration of the given solution in
the Schulz absorption cell, determine the value of θ' with the
observing tube T, Fig. 96, at six or eight different elevations.
Note the readings on the micrometer screw M. From the data
thus obtained, plot an absorption curve with micrometer scale
readings as abscissæ and $\tan^2 \theta$ as ordinates. This curve shows
the variation of absorption in different parts of the spectrum.
And inasmuch as that portion of the spectrum in which a given
change of wave-length produces the greatest change in absorption
is also the portion of the spectrum in which a given change of
concentration produces the greatest change in absorption, this is
the part of the spectrum to be used in the subsequent work on
this solution.

Make up a series of solutions of the given substance of known
concentrations. These may be expressed in per cent concentration, or in per cent normal concentration. With the observing
tube fixed at the elevation just found, find the value of θ' corresponding to each concentration. With these data plot a curve
coördinating concentration and $\log \tan \theta'$. Note the limits of
concentration within which the spectrophotometric method can be
used.

Select three or four convenient points on the straight portion
of the curve just plotted, compute the corresponding values of θ',
and plot a curve coördinating θ' and concentration. This is the
required standard spectrophotometric curve of the given solution. By its aid, an unknown concentration of the same solution
can be determined from a single reading of the instrument.

Exp. 47. Determination of the Concentration of a Solution by Means of a Dubosc Colorimeter

THEORY OF THE EXPERIMENT. — Read Art. 60. The object of this experiment is to make up a series of solutions having known concentrations of the same substance, to determine their relative concentrations by means of the Dubosc colorimeter, and to compare these calculated values with the known values.

MANIPULATION. — Copper sulphate is a convenient substance for this experiment. It should be remembered that if copper sulphate crystals be exposed to dry air the surface layer will lose part of the water of crystallization. In order that each molecule of copper sulphate may possess the full five molecules of water, it is best to recrystallize the salt and dry it between filter paper. For the purpose of this experiment, however, it will be sufficient to break up the large crystals of commercial copper sulphate and select for use the fragments from the interior of the crystals.

Make up 200 cc. of a molar solution of copper sulphate. Reserving enough of this solution to half fill one tube of the colorimeter, make up from the remainder a $\frac{8}{10}$ molar, a $\frac{6}{10}$ molar, a $\frac{4}{10}$ molar, and an $\frac{2}{10}$ molar solution. For diluting the molar solution to the required concentrations, use a 10-cc. pipette and a 100-cc. flask.

The scales beside the specimen tubes of colorimeters are usually divided into 100 equal spaces. If the instrument being used is divided in this manner, pour into the left-hand specimen tube T, Fig. 101, sufficient of the molar solution to come somewhat above the 10-division mark and lower the glass rod or closed tube T' till the attached index points to the 10-division mark. This adjustment need not be altered throughout the experiment.

Pour into the right-hand specimen tube one of the other solutions and adjust the thickness of the layer till the two halves of the field of view in the eyepiece are equally bright. Note the thickness of the layer.

Do the same for each of the other solutions.

Compute the concentration of each solution relative to the molar solution. Compare the results of the experiment with the known concentrations.

Exp. 48. A Spectrophotometric Comparison of Two Light Sources

THEORY OF THE EXPERIMENT. — Read Arts. 57 and 58.

In this experiment the ratio of the luminous intensities, at various wave-lengths, of a metal filament incandescent lamp and a carbon filament lamp are to be obtained by means of a Lemon-Brace spectrophotometer, and the results plotted in a curve.

MANIPULATION. — After putting the instrument into adjustment, calibrate it as described in Exp. 42. Place one of the light sources in front of one of the collimator slits, and the other source in front of the other slit. Rotate the telescope into a position in which the field of view is filled by light from one end of the spectrum. Note the angular position of the telescope. Rotate one Nicol prism till the two parts of the field of view are equally bright and note the angle between the planes of polarization of the two Nicols. Take a series of similar observations at five-degree intervals throughout the range of the visible spectrum.

Observe whether the part of the field of view due to light that has traversed the two Nicol prisms is maximum or minimum when the Nicols are set at zero. If the former, compute the relative intensities of the two sources by (160). If the latter, compute the relative intensities by (161). For the computation of relative intensities use Table 13.

Plot a curve having wave-lengths as abscissæ and relative intensities as ordinates.

Throughout the experiment the electric currents through the lamps must be constant. This constancy is most easily attained by operating the lamps by a storage battery.

Exp. 49. Determination of the Absorption of Light of Different Wave-Lengths produced by a Given Substance

THEORY OF THE EXPERIMENT. — Read Arts. 57 and 58. The object of this experiment is to determine the fraction of the light of different wave-lengths incident upon a given specimen that is transmitted by it.

MANIPULATION. — Read Exp. 48. Place a storage battery operated incandescent lamp in front of one of the collimator slits of a Lemon-Brace spectrophotometer. Arrange a mirror so that light from the same source traverses the other collimator. Place the specimen under investigation in front of the latter collimator slit, and make a series of observations as in the previous experiment.

Plot a curve having as abscissæ, wave-lengths; and as ordinates, the ratios of the light transmitted by, to the light incident on, the specimen.

Exp. 50. Determination of the Amount of Sucrose in a Sample of Sugar by Means of a Polarimeter

THEORY OF THE EXPERIMENT. — Read Arts. 61–68.

MANIPULATION. — A sodium flame is to be placed at such a point X, Fig. 103, that the image is formed in the plane of the diaphragm D_2. This is readily effected by fastening a needle to the lamp, just in front of the flame, and then moving the lamp back and forth till an image of the needle is formed on a white card held in front of the diaphragm D_2.

Fill the 2-decimeter specimen tube with distilled water and place it in the polarimeter. Rotate the analyzer till the position is found at which the entire field of view is uniformly bright. Note the scale reading. This is the zero setting of the apparatus.

Weigh out from 10 to 20 gm. of the specimen and place it in a graduated 100-cc. flask about half filled with distilled water. If the solution is clear and colorless, dilute it to the 100-cc. mark. But if the solution is not clear and colorless it must first be clarified as in Art. 68, and then diluted to the 100-cc. mark.

Empty the 2-decimeter solution tube, rinse it with some of the sugar solution, and then fill with the solution. Place the specimen tube in the polarimeter and note the temperature and the scale reading when the entire field of view is uniformly bright.

Take another flask graduated to both 50 cc. and 55 cc. and fill with the sugar solution to the 50-cc. mark. Add concentrated hydrochloric acid until the 55-cc. mark is reached. A small pipette is convenient in making these fine adjustments of volume. Pour the acidulated solution into a larger flask in a water bath so arranged that the temperature of the solution is maintained at between 70° C. and 80° C., for 10 minutes. At the end of this time, the sucrose in the solution is completely inverted. After the solution has cooled to room temperature, use a part of it to rinse out the 2.2-decimeter specimen tube. Then fill this tube with the invert sugar solution, place it in the polarimeter and note the temperature and the scale reading when the entire field of view is uniformly bright.

The data are now at hand for computing the amount of sucrose in the solution by means of (169). If a noninvertable sugar be present, its amount can afterward be computed by means of (170).

The following example will illustrate the method of computation.

Data. — Mass of sample of granulated sugar, 10.01 gm.

Zero point of polarimeter, *i.e.*, reading with specimen tube filled with distilled water, 6°.25.

Reading with 2-decimeter tube filled with solution of 10.01 gm. sugar in sufficient water to make 100 cc. of solution, 19°.55.

Reading with 2.2-decimeter tube filled with solution after inversion, 1°.85. Temperature 22° C.

Computation. — From the above data,

$$\theta = 19°.55 - 6°.25 = 13°.30.$$
$$\theta' = - (6°.25 - 1°.85) = -4°.40.$$

The values of the specific rotations of sucrose and invert sugar are given in Art. 61. Since ordinary granulated sugar is so nearly pure sucrose, we will introduce a negligible error in assuming that the concentration of sucrose in the specimen tube was 10 gm. per 100 cc. of solution, and that the concentration of invert sugar was 10 (1.05) gm. per 100 cc. of solution. Substituting these numbers in the values of the specific rotations of sucrose and invert sugar, Art. 61, we have for sucrose

$$[a_1]_t = 66.51 + 0.0045 (10) - 0.0144 (22 - 20) = 66°.53,$$

and for invert sugar

$$[a_2]_t = - 19.8 - 0.036 (10 \times 1.05) + 0.304 (22 - 20) = - 19°.57.$$

Substituting these values in (169), we obtain

$$c_1 = \frac{13.30 + 4.40}{2 (66.53 + 1.158 \times 19.57) 0.01}$$
$$= 9.92 \text{ gm. sucrose in 100 cc. of solution.}$$

And as the solution contained 10.01 gm. of the sample, the sample contained 99.1 per cent of sucrose.

TABLE 1.—CONVERSION FACTORS

LENGTH

1 centimeter = 0.39371 inch	1 inch = 2.53995 cm.
1 meter = 3.2809 feet	1 foot = 0.30479 m.
1 kilometer = 0.62138 mile	1 mile = 1.60931 Km.
1 micron = 0.001 mm.	1 mil = 0.001 inch
= 0.0000394 inch	= 0.00254 cm.

AREA

1 sq. cm. = 0.15501 sq. in.	1 sq. in. = 6.4514 sq. cm.
1 sq. m. = 10.764 sq. ft.	1 sq. ft. = 0.092900 sq. m.

VOLUME

1 cu. cm. = 0.061027 cu. in.	1 cu. in. = 16.386 cu. cm.
1 cu. m. = 35.317 cu. ft.	1 cu. ft. = 0.028315 cu. m.
1 liter = 1.76077 pints	1 quart = 1.13586 liters

MASS

1 gram = 15.43235 grains	1 grain = 0.064799 gram
= 5 carats (diamond)	1 dram (Adv.) = 1.772 grams
1 kilogram = 2.20462 lb.	1 lb. (7000 grs.) = 0.45359 Kg.

ANGLE

1 radian = 57.296 degrees	1 degree = 0.017453 radian

DENSITY

1 g. per cc.	1 lb. per cu. ft.
= 62.425 lb. per cu. ft.	= 0.016019 g. per cc.

FORCE

1 dyne = 0.000072331 poundal	1 poundal = 13825 dynes
1 g. wt. = 0.0022046 lb. wt.	1 lb. wt. = 453.59 g. wt.

MOMENT OF INERTIA

1 cm. g. unit	1 ft. lb. unit
= 2.3731 × 10⁻⁴ ft. lb. units	= 421390 cm. g. units

STRESS

1 dyne per sq. cm.
    = 0.067197 poundal per sq. ft.
1 g. wt. per sq. cm.
    = 2.0482 lb. wt. per sq. ft.
1 cm. of mercury at 0° C.
    = 13.596 g. wt. per sq. cm.
    = 0.19338 lb. wt. per sq. in.

'1 poundal per sq. ft.
    = 14.8816 dynes per sq. cm.
1 lb. wt. per sq. ft.
    = 0.48824 g. wt. per sq. cm.
1 in. of mercury at 0° C.
    = 34.533 g. wt. per sq. cm.
    = 0.49117 lb. wt. per sq. in.

WORK OR ENERGY

1 erg   = 2.3731×10^{-4} ft. poundals
1 joule = 10^7 ergs
         = 23.731 ft. poundals
1 g. cm. = 7.233×10^{-4} ft. lb.

1 ft. poundal = 421390 ergs
1 ft. lb.   = 13825.5 g. cm.
              = 1.35485 joules
1 h.p. hour   = 2685600 joules

POWER

1 watt = 10^7 ergs per sec.
       = 23.731 ft. poundals per sec.
       = 44.23 ft. pounds per min.
1 force de cheval
       = 75 Kg. m. per sec.
       = 0.9863 horse power

1 ft. poundal per sec.
       = 421390 ergs per sec.
1 ft. lb. per min.
       = 0.13825 Kg. m. per min.
1 horse power = 745.96 watts
              = 1.0139 force de cheval

THERMOMETRIC SCALES

$$C = \tfrac{5}{9}(F - 32) \qquad F = \tfrac{9}{5}C + 32$$

UNIT QUANTITY OF HEAT

1 g. calorie = 0.0039683 B.t.u. | 1 B.t.u. = 252.00 g. calories

MECHANICAL EQUIVALENT OF HEAT *

1 g. calorie = 4.19 joules
             = 426.9 Kg. m.
             = 1400.6 ft. lb.

1 B.t.u. = 1055 joules
         = 778.1 ft. lb.

LOGARITHMS

$\log_{10} N$ = $0.43429 \log_e N$ | $\log_e N = 2.3026 \log_{10} N$

* Computed with the value of g at Greenwich.

TABLE 2.—DENSITIES OF SOLIDS AND LIQUIDS

Since density varies with the temperature and with the specimen, these numbers are to be regarded as approximations only.

Substance	Grams per c.c.	Lbs. per cu. ft.	Substance	Grams per c.c.	Lbs. per cu. ft.
Aluminium	2.7	170	Lime	2.3	140
NH$_4$Cl	1.52	95		3.2	200
Antimony	6.71	419	Limestone	2.5	150
Asbestos	2.0	125		3.0	190
	2.8	175	Marble	2.6	160
Asphalt	1.0	62		2.8	175
	1.8	110	Mica	2.6	160
Beeswax	0.96	60		2.9	180
Benzene	0.70	44	Mercury at 0° C	13.596	848.7
Bismuth	9.80	612	Nickel	8.90	556
Brass	7.7	480	Oil, linseed	0.94	59
	8.7	540	Oil, olive	0.91	57
Brick	1.6	100	Paraffin	0.87	54
	2.1	130		0.93	58
Bronze	8.6	540	Phosphorus	1.83	114
CaCl$_2$	2.2	140	Platinum	21.5	1340
CS$_2$ at 20° C	1.264	78.9	Porcelain	2.4	150
Chalk	1.8	110	K$_2$CrO$_4$	2.72	170
	2.8	175	K$_2$Cr$_2$O$_7$	2.70	169
Coal	1.2	75	Quartz	2.65	165
	1.8	110	Resin	1.07	67
Copper	8.92	557	Sandstone	2.2	140
CuSO$_4$	2.27	142		2.5	150
Cork	0.24	15	Shellac	1.1	70
Diamond	3.52	220		1.2	75
Ether at 0° C	0.736	45.9	Silver { pure	10.53	657
German silver	8.62	538	{ mint	10.38	648
Glass	2.5	150	Slate	2.7	170
	3.9	250	Soapstone	2.7	170
Glycerin	1.26	79	Solder (soft)	8.9	555
Gold, pure	19.32	1206	NaCl	2.15	134
Granite	2.5	150	Sulphur, rhombic	2.07	129
	3.0	190	Tin	7.29	455
Graphite	2.3	140	Turpentine	0.87	54
Ice at 0° C	0.9167	57.22	Vulcanite	1.22	76
Iron cast	7.0	440	Water at 4° C	1.000013	62.4252
	7.7	480	Woods sea-soned ash	0.75	47
pure	7.86	491	cherry	0.67	42
Iron steel	7.6	470	oak	0.7	45
	7.8	490		1.0	62
wrought	7.79	486	pine	0.5	31
	7.85	490	poplar	0.4	25
Ivory	1.83	114	walnut	0.7	45
	1.92	120	Zinc	7.15	446
Lead (cast)	11.34	708	ZnSO$_4$	2.0	125

TABLE 3. — SPECIFIC GRAVITY OF WATER AT DIFFERENT TEMPERATURES

REFERRED TO WATER AT 4° C.

°C.	Sp. gr.	°C.	Sp. gr.	°C.	Sp. gr.	°C.	Sp. gr.	°C.	Sp. gr.
−4	0.99945	17	0.99882	38	0.99303	59	0.98382	80	0.97191
−3	58	18	864	39	268	60	331	81	129
−2	70	19	845	40	233	61	280	82	066
−1	79	20	825	41	195	62	228	83	004
0	87	21	804	42	157	63	175	84	0.96941
1	93	22	782	43	117	64	121	85	876
2	97	23	759	44	077	65	067	86	812
3	99	24	735	45	035	66	012	87	747
4	1.00000	25	710	46	0.98993	67	0.97957	88	682
5	0.99999	26	684	47	949	68	902	89	616
6	97	27	657	48	905	69	846	90	550
7	93	28	629	49	860	70	790	91	483
8	88	29	600	50	813	71	733	92	416
9	82	30	571	51	767	72	674	93	348
10	74	31	540	52	721	73	615	94	280
11	64	32	509	53	674	74	555	95	212
12	54	33	477	54	627	75	495	96	143
13	42	34	444	55	579	76	435	97	074
14	29	35	410	56	530	77	375	98	005
15	14	36	372	57	481	78	314	99	0.95934
16	0.99899	37	337	58	432	79	253	100	863

TABLE 4. — SPECIFIC GRAVITIES OF AQUEOUS SOLUTIONS OF ALCOHOL

Per cent alcohol by weight	Specific gravity at			Per cent alcohol by weight	Specific gravity at		
	10°	20°	30°		10°	20°	30°
0	0.99975	0.99831	0.99579	55	0.91074	0.90275	0.89456
5	0.99113	0.98945	0.98680	60	0.89944	0.89129	0.88304
10	0.98409	0.98195	0.97892	65	0.88790	0.87961	0.87125
15	0.97816	0.97327	0.97142	70	0.87613	0.86781	0.85925
20	0.97263	0.96877	0.96413	75	0.86427	0.85580	0.84719
25	0.96672	0.96185	0.95628	80	0.85215	0.84366	0.83483
30	0.95998	0.95403	0.94751	85	0.83967	0.83115	0.82232
35	0.95174	0.94514	0.93813	90	0.82665	0.81801	0.80918
40	0.94255	0.93511	0.92787	95	0.81291	0.80433	0.79553
45	0.93254	0.92493	0.91710	100	0.79788	0.78945	0.78096
50	0.95182	0.91400	0.90577				

TABLE 5. — SPECIFIC GRAVITIES OF AQUEOUS SOLUTIONS AT 15° C.

Referred to Water at 4° C.

Per cent	HCl	HNO₃	H₂SO₄	NaOH	NaCl	CuSO₄	ZnSO₄	Sugar at 17°.5	Per cent
0	0.9991	0.999	0.9991	0.999	0.999	0.999	0.999	0.9987	0
5	1.0242	1.029	1.0334	1.056	1.035	1.050	1.052	1.0184	5
10	1.0490	1.058	1.0687	1.111	1.072	1.103	1.108	1.0388	10
15	1.0744	1.089	1.1048	1.166	1.110	1.161	1.168	1.0600	15
20	1.1001	1.121	1.1430	1.222	1.150	1.225	1.236	1.0819	20
25	1.1262	1.154	1.1816	1.277	1.191	.....	1.307	1.1047	25
30	1.1524	1.187	1.223	1.133	.....	.....	1.382	1.1282	30
35	1.1775	1.220	1.264	1.387	.....	.....	.....	1.1526	35
40	1.2007	1.253	1.307	1.442	.....	.....	.....	1.1780	40
45	.....	1.287	1.352	1.496	.....	.....	.....	1.2041	45
50	.....	1.320	1.399	1.548	.....	.....	.....	1.2313	50
55	.....	1.350	1.449	.....	.....	.....	.....	1.2593	55
60	.....	1.377	1.503	.....	.....	.....	.....	1.2883	60
65	.....	1.402	1.559	.....	.....	.....	.....	1.3183	65
70	.....	1.424	1.616	.....	.....	.....	.....	1.3494	70
75	.....	1.443	1.675	.....	.....	.....	.....	1.3813	75
80	.....	1.461	1.733	.....	.....	.....	.....	.....	80
85	.....	1.470	1.785	.....	.....	.....	.....	.....	85
90	.....	1.497	1.819	.....	.....	.....	.....	.....	90
95	.....	1.514	1.839	.....	.....	.....	.....	.....	95
100	.....	1.530	1.838	.....	.....	.....	.....	.....	100

TABLE 6. — REDUCTION OF ARBITRARY HYDROMETER SCALES

Light liquids			Scale reading	Heavy liquids			
Baumé	Beck	Cartier		Baumé *	Baumé †	Beck	Twaddell
Sp. gr.	Sp. gr.	Sp. gr.		Sp. gr.	Sp. gr.	Sp. gr.	Sp. gr.
. . . .	1.000	. . . . . .	0	1.000	1.000	1.000	1.000
. . . .	0.971	. . . . . .	5	1.035	1.036	1.030	1.025
1.000	0.944	. . . . . .	10	1.073	1.074	1.062	1.050
0.967	0.919	0.970	15	1.114	1.116	1.097	1.075
0.936	0.895	0.936	20	1.158	1.161	1.133	1.100
0.907	0.872	0.905	25	1.205	1.210	1.172	1.125
0.880	0.850	0.876	30	1.257	1.262	1.214	1.150
0.854	0.829	0.849	35	1.313	1.320	1.259	1.175
0.830	0.810	0.824	40	1.375	1.384	1.308	1.200
0.807	0.791	. . . . . .	45	1.442	1.453	1.360	1.225
0.785	0.773	. . . . . .	50	1.517	1.530	1.417	1.250
0.764	0.756	. . . . . .	55	1.599	1.616	1.478	1.275
0.745	0.739	. . . . . .	60	1.691	1.712	1.545	1.300
. . . .	0.723	. . . . . .	65	1.795	1.820	1.619	1.325
. . . .	0.708	. . . . . .	70	1.912	1.920	1.700	1.350

* Original scale for liquids denser than water.   † Newer or so-called " rational " scale.

TABLE 7. — SPECIFIC GRAVITIES OF GASES AND VAPORS

REFERRED TO WATER AT 4° C.; ALSO TO AIR AND HYDROGEN AT 0° C. AND 760 MM. OF MERCURY PRESSURE

All results are given for a pressure of 760 mm. of mercury

Substance	Formula	Tempera-ture ° C.	Specific gravity referred to		
			Water	Air	Hydrogen
Air.................	....	0	0.0012931	1.0000	14.445
Ammonia............	NH_3........	0	0.0007616	0.5890	8.508
Carbon dioxide.......	CO_2........	0	0.001965	1.520	21.955
Chlorine......	Cl_2........	0	0.0031674	2.4500	35.382
Coal gas.............		0	0.000421	0.3256	4.715
		0	0.000667	0.5158	7.452
Hydrogen............	H_2........	0	0.0000895	0.0692	1.000
Nitrogen............	N_2........	0	0.0012546	0.9701	14.013
Oxygen..............	O_2........	0	0.0014292	1.1052	15.964
Acetic acid..........	CH_3COOH	125	0.00414	3.2	46.2
		250	0.00269	2.08	30.0
Amyl bromide.......	$C_5H_{11}Br$....	152	0.00703	5.43	78.5
		196	0.00604	4.67	67.5
		295	0.00412	3.18	46.0
		360	0.00340	2.63	38.0
Ammonium chloride *	NH_4Cl.....	300	0.00128	0.986	14.23
		360	0.00122	0.944	13.63
		448	0.00120	0.932	13.45
Iodine..............	I_2.........	448	0.01130	8.74	126.9
		680	0.01064	8.23	118.8
		855	0.01043	8.07	116.5
		1043	0.00906	7.01	101.2
		1275	0.00753	5.82	84.0
		1468	0.00657	5.06	73.0
Nitrogen peroxide....	N_2O_4......	4.2	0.00335	2.588	37.36
		49.6	0.00294	2.27	32.77
		60.2	0.00269	2.08	30.03
		70.0	0.00248	1.92	27.72
		90.0	0.00222	1.72	24.83
		100.1	0.00217	1.68	24.25
		154.0	0.00204	1.58	22.81

* Ammonium chloride vapor gives abnormal vapor densities only when in presence of moisture

TABLE 8. — COEFFICIENTS OF FRICTION

Substance	Static coefficient μ	Kinetic coefficient b
	from	from
Metals on metals (dry)............	0.2 to 0.4	0.18 to 0.35
Metals on metals (wet)............	0.15 to 0.3	0.14 to 0.28
Metals on metals (oiled)..........	0.15 to 0.2	0.14 to 0.18
Wood on wood (dry)*..............	0.5 to 0.7	0.2 to 0.3
Wood on wood (dry) †.............	0.4 to 0.6	0.18 to 0.3
Leather belt on wood pulley........	0.45 to 0.6	0.3 to 0.5
Leather belt on iron pulley.........	0.25 to 0.35	0.2 to 0.3

* Motion in direction of fiber.   † Motion normal to fiber of sliding block.

TABLE 9. — ELASTIC CONSTANTS OF SOLIDS

N.B. — Flexural Resilience per unit volume equals one-ninth the Tensile Resilience per unit volume

Substance	Young's modulus		Elastic limit		Breaking stress		Simple rigidity		Tensile resilience	
	dynes sq.cm.	lbs. sq. in.	dynes sq.cm	lbs. sq. in.	dynes sq.cm.	lbs. sq. in.	dynes sq.cm.	lbs. sq. in.	ergs cu. cm.	ft. lbs. cu. ft.
Multiply by ..	10^{11}	10^6	10^6	10^3	10^6	10^3	10^{11}	10^6	10^4	1
BRASS:										
cast........	6.5	9	4.5	6	20	30	2.4	3.5	16	330
wire........	10	14	11	16	60	80	3.7	5.4	60	1300
COPPER:										
annealed....	10	14	3	4	31	43	...	...	5	190
cast........	12	17	4.5	6.3	18	25	4.0	6.0	8	170
wire........	12	17	7	10	40	55	4.5	6.5	20	420
GLASS	6.5	9	2.3	3.2	2–9	3–12	2.4	3.5	4	80
IRON:										
annealed....	21	30	5	7	50	70	...	...	6	130
cast........	12	17	7	10	15	20	5.3	7.6	20	420
wire........	19	26	20	30	60	85	8.0	12.0	100	2000
wrought....	20	28	20	30	40	55	7.7	11.0	100	2000
STEEL:										
Bessemer...	22	31	33	46	70	100	...	...	250	5200
cast........	20	28	...	...	40	60	8.0	12.0	‡5600	‡120000
hearth......	21	30	...	...	70	100	...	...	......	......
wire........	19	26	*40	*60	110	150	...	...	*420	*8800
WOODS:										
oak........	1.0	1.4	2.3	3.2	†5	†7	...	...	27	560
pine........	1.1	1.6	2.4	3.3	†4	†5	...	...	26	540
poplar......	0.5	0.7	1.5	2.2	†3	†4	...	...	23	480

* Unannealed.   † Parallel to grain.   ‡ Hardened.

TABLE 10. — VISCOSITIES OF WATER AND AQUEOUS SUGAR SOLUTIONS

η denotes the coefficient of viscosity in c.g.s. units, z_0, z_{20}, etc., the specific viscosity, or viscosity relative to water at 0° C., 20° C., etc.

(a) WATER AT DIFFERENT TEMPERATURES

Temp.	η	z_0	Temp.	η	z_0
0°	0.01809	1.000	30°	0.00812	0.449
5	0.01530	0.846	40	0.00664	0.367
10	0.01326	0.733	50	0.00570	0.315
15	0.01150	0.636	60	0.00487	0.269
20	0.01016	0.562	70	0.00424	0.235
25	0.00903	0.499	....	....	....

(b) AQUEOUS SOLUTIONS OF SUGAR OF VARIOUS CONCENTRATIONS AT 20° C.

Per cent sugar	z_{20}	Per cent sugar	z_{20}	Per cent sugar	z_{20}
2	1.0521	12	1.4110	22	2.0552
4	1.1104	14	1.5092	24	2.2454
6	1.1840	16	1.6196	26	2.4540
8	1.2576	18	1.7484	28	2.7055
10	1.3312	20	1.8895	30	3.0674

TABLE 11. — THE GREEK ALPHABET

Letter	Name	Letter	Name	Letter	Name
A, α	Alpha	I, ι	Iota	P, ρ	Rho
B, β	Beta	K, κ	Kappa	Σ, σ	Sigma
Γ, γ	Gamma	Λ, λ	Lambda	T, τ	Tau
Δ, δ	Delta	M, μ	Mu	Y, υ	Upsilon
E, ϵ	Epsilon	N, ν	Nu	Φ, ϕ	Phi
Z, ζ	Zeta	Ξ, ξ	Xi	X, χ	Chi
H, η	Eta	O, o	Omicron	Ψ, ψ	Psi
Θ, θ	Theta	Π, π	Pi	Ω, ω	Omega

TABLE 12. — FACTORS BY WHICH THE VOLUME OF A FIXED
BAROMETRIC PRESSURE (IN INCHES OF MERCURY), AND
IN ORDER TO OBTAIN THE VOLUME THE MASS WOULD

Pressure in inches of mercury	Temperature										
	40° F.	42° F.	44° F.	46° F.	48° F.	50° F.	52° F.	54° F.	56° F.	58° F.	60° F.
28.0	0.979	0.974	0.970	0.965	0.960	0.956	0.951	0.946	0.942	0.937	0.932
28.1	0.983	0.978	0.973	0.969	0.964	0.959	0.955	0.951	0.945	0.941	0.936
28.2	0.986	0.981	0.977	0.972	0.967	0.963	0.958	0.953	0.949	0.944	0.939
28.3	0.990	0.985	0.980	0.976	0.971	0.966	0.961	0.957	0.952	0.947	0.942
28.4	0.993	0.988	0.984	0.979	0.974	0.970	0.965	0.960	0.955	0.951	0.946
28.5	0.997	0.992	0.987	0.983	0.978	0.973	0.968	0.964	0.959	0.954	0.949
28.6	1.001	0.995	0.991	0.986	0.981	0.977	0.972	0.967	0.962	0.958	0.953
28.7	1.004	0.999	0.994	0.990	0.985	0.980	0.975	0.970	0.966	0.961	0.956
28.8	1.007	1.003	0.998	0.993	0.988	0.984	0.979	0.974	0.969	0.964	0.959
28.9	1.011	1.006	1.001	0.997	0.992	0.987	0.982	0.977	0.973	0.968	0.963
29.0	1.014	1.010	1.005	1.000	0.995	0.990	0.986	0.981	0.976	0.971	0.966
29.1	1.018	1.013	1.008	1.004	0.999	0.994	0.989	0.984	0.979	0.975	0.969
29.2	1.021	1.017	1.012	1.007	1.002	0.997	0.992	0.988	0.982	0.978	0.973
29.3	1.025	1.020	1.015	1.011	1.006	1.001	0.996	0.991	0.986	0.981	0.976
29.4	1.028	1.024	1.019	1.014	1.009	1.004	0.999	0.995	0.990	0.985	0.980
29.5	1.032	1.027	1.022	1.018	1.013	1.008	1.003	0.998	0.993	0.988	0.983
29.6	1.036	1.031	1.026	1.021	1.016	1.011	1.006	1.001	0.996	0.992	0.986
29.7	1.039	1.034	1.029	1.025	1.019	1.015	1.010	1.005	1.000	0.995	0.990
29.8	1.043	1.038	1.033	1.028	1.023	1.018	1.013	1.008	1.003	0.998	0.993
29.9	1.046	1.041	1.036	1.031	1.026	1.022	1.017	1.012	1.007	1.002	0.997
30.0	1.050	1.045	1.040	1.035	1.030	1.025	1.020	1.015	1.010	1.005	1.000
30.1	1.053	1.048	1.043	1.038	1.033	1.029	1.024	1.019	1.014	1.009	1.003
30.2	1.057	1.052	1.047	1.042	1.037	1.032	1.027	1.022	1.017	1.012	1.007
30.3	1.060	1.055	1.050	1.045	1.040	1.036	1.030	1.024	1.020	1.015	1.010
30.4	1.064	1.059	1.054	1.049	1.044	1.039	1.034	1.029	1.024	1.019	1.014
30.5	1.067	1.062	1.057	1.052	1.047	1.042	1.037	1.032	1.027	1.022	1.017
30.6	1.071	1.066	1.061	1.056	1.051	1.046	1.041	1.036	1.031	1.026	1.020
30.7	1.074	1.069	1.064	1.059	1.054	1.049	1.044	1.039	1.034	1.029	1.024
30.8	1.078	1.073	1.068	1.063	1.058	1.053	1.048	1.043	1.037	1.032	1.027
30.9	1.081	1.076	1.071	1.066	1.061	1.056	1.051	1.046	1.041	1.036	1.031
31.0	1.085	1.080	1.075	1.070	1.065	1.060	1.055	1.049	1.044	1.039	1.034

MASS OF GAS SATURATED WITH WATER VAPOR, AT A GIVEN
AT A GIVEN TEMPERATURE (IN ° F.), MUST BE MULTIPLIED,
HAVE AT 30 INCHES OF MERCURY AND 60° F.

Pressure in inches of mercury	Temperature											
	62° F.	64° F.	66° F.	68° F.	70° F.	72° F.	74° F.	76° F.	78° F.	80° F.	82° F.	84° F.
28.0	0.927	0.922	0.917	0.912	0.907	0.902	0.897	0.892	0.887	0.881	0.875	0.870
28.1	0.930	0.926	0.921	0.916	0.911	0.905	0.900	0.895	0.890	0.884	0.879	0.873
28.2	0.934	0.929	0.924	0.919	0.914	0.909	0.904	0.898	0.893	0.887	0.882	0.876
28.3	0.937	0.932	0.928	0.922	0.917	0.912	0.907	0.902	0.896	0.891	0.885	0.880
28.4	0.941	0.936	0.931	0.926	0.921	0.915	0.910	0.905	0.900	0.894	0.888	0.883
28.5	0.944	0.939	0.934	0.929	0.924	0.919	0.914	0.908	0.903	0.897	0.892	0.886
28.6	0.947	0.943	0.938	0.932	0.927	0.922	0.917	0.912	0.906	0.901	0.895	0.889
28.7	0.951	0.946	0.941	0.936	0.931	0.925	0.920	0.915	0.909	0.904	0.898	0.893
28.8	0.954	0.949	0.944	0.939	0.934	0.929	0.924	0.918	0.913	0.907	0.901	0.896
28.9	0.958	0.953	0.948	0.942	0.937	0.932	0.927	0.921	0.916	0.910	0.905	0.899
29.0	0.961	0.956	0.951	0.946	0.941	0.935	0.930	0.925	0.919	0.914	0.908	0.903
29.1	0.964	0.959	0.954	0.949	0.944	0.939	0.933	0.928	0.923	0.917	0.911	0.906
29.2	0.968	0.963	0.958	0.952	0.947	0.942	0.937	0.931	0.926	0.920	0.914	0.909
29.3	0.971	0.966	0.961	0.956	0.950	0.945	0.940	0.935	0.929	0.923	0.918	0.912
29.4	0.975	0.969	0.964	0.959	0.954	0.949	0.943	0.938	0.932	0.927	0.921	0.915
29.5	0.978	0.973	0.968	0.962	0.957	0.952	0.947	0.941	0.936	0.930	0.924	0.919
29.6	0.981	0.976	0.971	0.966	0.960	0.955	0.950	0.944	0.939	0.933	0.927	0.922
29.7	0.985	0.980	0.974	0.969	0.964	0.959	0.953	0.948	0.942	0.937	0.931	0.925
29.8	0.988	0.983	0.978	0.972	0.967	0.962	0.957	0.951	0.946	0.940	0.934	0.928
29.9	0.991	0.986	0.981	0.976	0.970	0.965	0.960	0.954	0.949	0.943	0.937	0.932
30.0	0.995	0.990	0.985	0.979	0.974	0.968	0.963	0.958	0.952	0.946	0.941	0.935
30.1	0.998	0.993	0.988	0.983	0.977	0.972	0.966	0.961	0.955	0.950	0.944	0.938
30.2	1.002	0.996	0.991	0.986	0.980	0.975	0.970	0.964	0.959	0.953	0.947	0.941
30.3	1.005	1.000	0.995	0.989	0.984	0.978	0.973	0.968	0.962	0.956	0.950	0.945
30.4	1.008	1.003	0.998	0.993	0.987	0.982	0.976	0.971	0.965	0.959	0.954	0.948
30.5	1.012	1.006	1.001	0.996	0.990	0.985	0.980	0.974	0.969	0.963	0.957	0.951
30.6	1.015	1.010	1.005	0.999	0.994	0.988	0.983	0.977	0.972	0.966	0.960	0.954
30.7	1.018	1.013	1.008	1.003	0.997	0.992	0.986	0.981	0.975	0.969	0.963	0.957
30.8	1.022	1.017	1.011	1.006	1.000	0.995	0.990	0.984	0.978	0.972	0.967	0.961
30.9	1.025	1.020	1.015	1.009	1.004	0.998	0.993	0.987	0.982	0.976	0.970	0.964
31.0	1.029	1.023	1.018	1.013	1.007	1.002	0.996	0.991	0.985	0.979	0.973	0.967

TABLE 13. — VALUES OF

The angles whose sines squared are given below are indicated in the
unity minus the sine squared of that angle.

θ°	0	0.1	0.2	0.3	0.4	0.5	0.6	0.7	0.8	0.9
0	0.0000	0.0000	0.0000	0.0000	0.0000	0.0000	0.0001	0.0001	0.0002	0.0002
1	0.0003	0.0004	0.0004	0.0005	0.0006	0.0007	0.0008	0.0009	0.0010	0.0011
2	0.0012	0.0013	0.0015	0.0016	0.0018	0.0019	0.0021	0.0022	0.0024	0.0026
3	0.0027	0.0029	0.0031	0.0033	0.0035	0.0037	0.0039	0.0042	0.0044	0.0046
4	0.0049	0.0051	0.0054	0.0056	0.0059	0.0062	0.0064	0.0067	0.0070	0.0073
5	0.0076	0.0079	0.0082	0.0085	0.0089	0.0092	0.0095	0.0099	0.0102	0.0106
6	0.0109	0.0113	0.0117	0.0120	0.0124	0.0128	0.0132	0.0136	0.0140	0.0144
7	0.0149	0.0153	0.0157	0.0161	0.0166	0.0170	0.0175	0.0180	0.0184	0.0189
8	0.0194	0.0199	0.0203	0.0208	0.0213	0.0218	0.0224	0.0229	0.0234	0.0239
9	0.0245	0.0250	0.0256	0.0261	0.0267	0.0272	0.0278	0.0284	0.0290	0.0296
10	0.0302	0.0307	0.0314	0.0320	0.0326	0.0332	0.0339	0.0345	0.0351	0.0358
11	0.0364	0.0371	0.0377	0.0384	0.0391	0.0398	0.0404	0.0411	0.0418	0.0425
12	0.0432	0.0439	0.0447	0.0454	0.0461	0.0469	0.0476	0.0483	0.0491	0.0499
13	0.0506	0.0514	0.0522	0.0529	0.0537	0.0545	0.0553	0.0561	0.0569	0.0577
14	0.0585	0.0594	0.0602	0.0610	0.0619	0.0627	0.0635	0.0644	0.0653	0.0662
15	0.0670	0.0679	0.0687	0.0696	0.0705	0.0714	0.0723	0.0732	0.0741	0.0751
16	0.0760	0.0769	0.0778	0.0788	0.0797	0.0807	0.0816	0.0826	0.0835	0.0845
17	0.0855	0.0864	0.0875	0.0884	0.0894	0.0904	0.0914	0.0924	0.0935	0.0945
18	0.0955	0.0965	0.0975	0.0986	0.0997	0.1007	0.1017	0.1028	0.1039	0.1040
19	0.1060	0.1071	0.1081	0.1092	0.1103	0.1115	0.1125	0.1137	0.1148	0.1159
20	0.1170	0.1181	0.1192	0.1203	0.1215	0.1226	0.1238	0.1250	0.1261	0.1272
21	0.1284	0.1296	0.1308	0.1320	0.1331	0.1344	0.1355	0.1367	0.1379	0.1391
22	0.1404	0.1415	0.1428	0.1440	0.1452	0.1464	0.1477	0.1490	0.1502	0.1514
23	0.1527	0.1540	0.1552	0.1565	0.1578	0.1590	0.1603	0.1616	0.1629	0.1641
24	0.1654	0.1667	0.1680	0.1694	0.1707	0.1720	0.1733	0.1746	0.1760	0.1773
25	0.1786	0.1800	0.1813	0.1826	0.1840	0.1854	0.1867	0.1880	0.1894	0.1908
26	0.1921	0.1936	0.1949	0.1963	0.1977	0.1991	0.2005	0.2020	0.2033	0.2047
27	0.2061	0.2075	0.2089	0.2104	0.2117	0.2132	0.2147	0.2161	0.2175	0.2190
28	0.2204	0.2218	0.2233	0.2248	0.2263	0.2277	0.2292	0.2306	0.2321	0.2336
29	0.2351	0.2365	0.2380	0.2394	0.2410	0.2424	0.2440	0.2455	0.2469	0.2485
30	0.2500	0.2515	0.2530	0.2546	0.2561	0.2576	0.2592	0.2606	0.2622	0.2638
31	0.2652	0.2668	0.2684	0.2699	0.2714	0.2730	0.2745	0.2760	0.2777	0.2793
32	0.2809	0.2824	0.2839	0.2855	0.2871	0.2887	0.2903	0.2919	0.2935	0.2950
33	0.2966	0.2983	0.2998	0.3015	0.3030	0.3046	0.3063	0.3079	0.3095	0.3110
34	0.3128	0.3143	0.3159	0.3175	0.3192	0.3208	0.3224	0.3240	0.3257	0.3272
35	0.3290	0.3307	0.3322	0.3339	0.3356	0.3373	0.3389	0.3406	0.3421	0.3439
36	0.3455	0.3472	0.3488	0.3504	0.3522	0.3538	0.3552	0.3571	0.3588	0.3606
37	0.3622	0.3639	0.3656	0.3673	0.3690	0.3705	0.3722	0.3739	0.3757	0.3774
38	0.3790	0.3807	0.3825	0.3841	0.3858	0.3874	0.3892	0.3908	0.3926	0.3943
39	0.3961	0.3978	0.3994	0.4012	0.4029	0.4046	0.4063	0.4079	0.4098	0.4115
40	0.4133	0.4150	0.4167	0.4184	0.4202	0.4217	0.4235	0.4252	0.4270	0.4288
41	0.4303	0.4321	0.4339	0.4355	0.4373	0.4391	0.4408	0.4426	0.4442	0.4461
42	0.4477	0.4496	0.4512	0.4529	0.4548	0.4565	0.4582	0.4598	0.4617	0.4634
43	0.4651	0.4669	0.4686	0.4703	0.4721	0.4738	0.4756	0.4773	0.4791	0.4808
44	0.4826	0.4844	0.4860	0.4878	0.4896	0.4914	0.4930	0.4948	0.4966	0.4982

SIN² θ AND COS² θ

first column and first line. The cosine squared of any angle equals

θ°	0	0.1	0.2	0.3	0.4	0.5	0.6	0.7	0.8	0.9
45	0.5000	0.5017	0.5035	0.5051	0.5070	0.5086	0.5105	0.5122	0.5141	0.5157
46	0.5174	0.5193	0.5210	0.5226	0.5243	0.5263	0.5280	0.5297	0.5314	0.5331
47	0.5351	0.5365	0.5383	0.5400	0.5418	0.5435	0.5453	0.5470	0.5488	0.5506
48	0.5523	0.5541	0.5557	0.5575	0.5593	0.5611	0.5626	0.5644	0.5663	0.5678
49	0.5696	0.5712	0.5731	0.5746	0.5765	0.5781	0.5800	0.5816	0.5834	0.5851
50	0.5869	0.5886	0.5902	0.5921	0.5937	0.5954	0.5970	0.5990	0.6006	0.6023
51	0.6046	0.6056	0.6073	0.6090	0.6107	0.6124	0.6140	0.6157	0.6174	0.6192
52	0.6209	0.6226	0.6243	0.6260	0.6280	0.6295	0.6310	0.6327	0.6345	0.6362
53	0.6377	0.6394	0.6412	0.6430	0.6445	0.6462	0.6477	0.6495	0.6513	0.6528
54	0.6546	0.6561	0.6580	0.6595	0.6610	0.6628	0.6644	0.6662	0.6677	0.6693
55	0.6710	0.6727	0.6742	0.6759	0.6775	0.6792	0.6808	0.6823	0.6841	0.6856
56	0.6872	0.6890	0.6906	0.6922	0.6937	0.6953	0.6970	0.6986	0.7002	0.7018
57	0.7034	0.7050	0.7065	0.7081	0.7097	0.7114	0.7129	0.7145	0.7160	0.7176
58	0.7191	0.7208	0.7223	0.7238	0.7254	0.7270	0.7284	0.7301	0.7316	0.7332
59	0.7347	0.7362	0.7377	0.7393	0.7408	0.7423	0.7439	0.7454	0.7470	0.7485
60	0.7501	0.7514	0.7530	0.7544	0.7560	0.7575	0.7589	0.7605	0.7621	0.7635
61	0.7649	0.7665	0.7679	0.7693	0.7707	0.7723	0.7737	0.7752	0.7768	0.7782
62	0.7796	0.7811	0.7825	0.7840	0.7854	0.7869	0.7881	0.7896	0.7910	0.7925
63	0.7940	0.7952	0.7967	0.7982	0.7995	0.8009	0.8022	0.8037	0.8050	0.8065
64	0.8078	0.8093	0.8106	0.8119	0.8134	0.8147	0.8160	0.8173	0.8187	0.8200
65	0.8215	0.8228	0.8241	0.8255	0.8268	0.8279	0.8293	0.8306	0.8320	0.8333
66	0.8346	0.8360	0.8372	0.8385	0.8397	0.8410	0.8424	0.8435	0.8449	0.8461
67	0.8474	0.8486	0.8498	0.8511	0.8523	0.8535	0.8549	0.8561	0.8572	0.8584
68	0.8596	0.8608	0.8622	0.8634	0.8646	0.8658	0.8670	0.8680	0.8692	0.8704
69	0.8716	0.8728	0.8740	0.8750	0.8762	0.8774	0.8784	0.8796	0.8808	0.8819
70	0.8831	0.8841	0.8853	0.8863	0.8876	0.8886	0.8896	0.8908	0.8919	0.8929
71	0.8939	0.8952	0.8962	0.8972	0.8983	0.8993	0.9003	0.9014	0.9024	0.9034
72	0.9045	0.9055	0.9066	0.9076	0.9087	0.9095	0.9105	0.9116	0.9126	0.9135
73	0.9145	0.9156	0.9164	0.9175	0.9183	0.9194	0.9203	0.9213	0.9221	0.9230
74	0.9241	0.9249	0.9258	0.9268	0.9277	0.9285	0.9294	0.9305	0.9313	0.9322
75	0.9330	0.9339	0.9348	0.9356	0.9365	0.9374	0.9382	0.9391	0.9397	0.9406
76	0.9415	0.9423	0.9432	0.9438	0.9447	0.9456	0.9463	0.9471	0.9478	0.9486
77	0.9493	0.9502	0.9508	0.9517	0.9524	0.9532	0.9539	0.9546	0.9554	0.9561
78	0.9568	0.9574	0.9581	0.9590	0.9596	0.9603	0.9609	0.9616	0.9623	0.9629
79	0.9636	0.9643	0.9650	0.9656	0.9660	0.9667	0.9674	0.9681	0.9687	0.9692
80	0.9698	0.9704	0.9710	0.9716	0.9722	0.9728	0.9733	0.9739	0.9744	0.9750
81	0.9755	0.9761	0.9766	0.9771	0.9777	0.9782	0.9787	0.9792	0.9797	0.9801
82	0.9806	0.9811	0.9816	0.9820	0.9825	0.9830	0.9834	0.9838	0.9843	0.9847
83	0.9852	0.9856	0.9860	0.9864	0.9868	0.9872	0.9876	0.9880	0.9883	0.9887
84	0.9891	0.9894	0.9898	0.9901	0.9905	0.9908	0.9912	0.9915	0.9918	0.9921
85	0.9924	0.9927	0.9930	0.9933	0.9936	0.9938	0.9941	0.9944	0.9946	0.9949
86	0.9951	0.9954	0.9956	0.9958	0.9960	0.9963	0.9965	0.9967	0.9969	0.9971
87	0.9973	0.9974	0.9976	0.9978	0.9979	0.9981	0.9983	0.9984	0.9985	0.9987
88	0.9988	0.9989	0.9990	0.9991	0.9992	0.9993	0.9994	0.9995	0.9996	0.9996
89	0.9997	0.9998	0.9998	0.9999	0.9999	0.9999	1.0000	1.0000	1.0000	1.0000

TABLE 14. — ABSOLUTE INDEX OF REFRACTION OF VARIOUS SUBSTANCES, FOR THE D LINE, $\lambda = 5893$ Å

Values for gases are for a temperature of 0° C. and pressure of 76 cm. of mercury. For other substances, values are for a temperature of 20° C., unless otherwise stated.

Air	1.0002926	Calcspar (ord. ray)	1.658
Carbon dioxide	1.00045	(exord. ray)	1.486
Water vapor	1.00025	Diamond	2.43
Canada balsam	1.54	Fluorite	1.434
Cassia oil	1.603	Ice (at −8° C.) (ord. ray)	1.309
Carbon disulphide	1.618	(exord. ray)	1.313
Cinnamon oil	1.619	Rocksalt	1.544
Ethyl alcohol	1.361	Quartz (ord. ray)	1.539
Water	1.333	(exord. ray)	1.548

TABLE 15. — INDEX OF REFRACTION AND DISPERSION OF OPTICAL GLASSES

The second column gives the index of refraction for the D line, and the third column gives the dispersion between the C and the F Fraunhofer lines.

	n_D	$(n_F - n_C)$
Light phosphate crown glass	1.5159	0.00737
Barium-silicate crown glass	1.5399	0.00909
High-dispersion crown glass	1.5262	0.01026
Borate flint glass	1.5686	0.01102
Extra light flint glass	1.5398	0.01142
Heavy flint glass	1.7174	0.02434
Heaviest flint glass	1.9626	0.04882

TABLE 16. — WAVE-LENGTHS OF THE PROMINENT LINES OF THE VISIBLE SOLAR SPECTRUM

At 15° C. and 76 cm. mercury pressure

The letters under the heading " Line " are the designations used by Fraunhofer.

Line	Due to	Wave-lengths in Ångström units	Line	Due to	Wave-lengths in Ångström units
A	O	7593.8	b_4	Mg	5167.3
B	O	6867.2	F	H	4861.4
C	H	6562.8	G	H	4340.5
D_1	Na	5895.9	G	{ Ca	4307.7
D_2	Na	5890.0		} Fe	4307.9
E	Fe	5269.6	h	H	4101.9
b_1	Mg	5183.6	H	Ca	3968.5
b_2	Mg	5172.7	K	Ca	3933.7

INDEX TO VOL. I